AF588325

T-Labs Series in Telecommunication Services

For further volumes:
http://www.springer.com/series/10013

Błażej Lewcio

Management of Speech and Video Telephony Quality in Heterogeneous Wireless Networks

Springer

Błażej Lewcio
Quality and Usability Lab
Telekom Innovation Laboratories
TU Berlin
Berlin
Germany

ISSN 2192-2810 ISSN 2192-2829 (electronic)
ISBN 978-3-319-02101-0 ISBN 978-3-319-02102-7 (eBook)
DOI 10.1007/978-3-319-02102-7
Springer Cham Heidelberg New York Dordrecht London

Library of Congress Control Number: 2013947796

Printed on acid-free paper

Springer is part of Springer Science+Business Media (www.springer.com)

Preface

This book shows how networking research and quality engineering can be combined to successfully manage the transmission quality when speech and video telephony is delivered in heterogeneous wireless networks. Nomadic use of services requires intelligent management of ongoing transmission, and to make the best of available resources many fundamental trade-offs must be considered. Network coverage versus throughput and reliability of a connection is one key aspect, efficiency versus robustness of signal compression is another. However, to successfully manage services, user-perceived Quality of Experience (QoE) in heterogeneous networks must be known, and the perception of quality changes must be understood. These issues are addressed in this book, in particular focusing on the perception of quality changes due to switching between diverse networks, speech and video codecs, and encoding bit rates during active calls.

To tackle the addressed issues, the related work is reviewed first in this book. Afterwards the research testbed that was created in scope of this work is described, and the results of numerous experiments are evaluated. During the analysis two primary goals are pursued for both speech and video telephony. First, user perception of the addressed transmission phenomena is studied and guidelines for perception-aware management of quality are derived. Second, a number of existing speech and video quality prediction models is validated in the context of heterogeneous wireless networks. The strengths and weaknesses of the individual methods are discussed, and some new quality modelling approaches are proposed.

It must be noted that this book is a result of myö doctoral research that would not have been possible without the support of numerous people. Foremost, I would like to thank Prof. Dr. -Ing. Sebastian Möller for his excellent guidance and immeasurable effort in supervising my research. I want to thank him for all the constructive comments and suggestions that helped me to improve the quality of my research. Also, I would like to thank him for providing excellent working conditions, and for countless opportunities to exchange my knowledge with top researchers from all over the world.

I would also like to thank Dr. Pablo Vidales, who introduced me to the topic of perception-based mobility management, and who provided me with excellent guidance when I made my first step in the international research community.

Last but not least, I would like to collectively thank all the people who made this research such an enjoyable and memorable experience. I would like to thank all the members of the Mobisense project, as well as all my colleagues from the Quality and Usability Lab (Q&U Lab) at Technische Universität Berlin. Finally, I would like to thank the management of Telekom Innovation Laboratories (T-Labs) for supporting this research.

Berlin, April 2013 Błażej Lewcio

Contents

Abbreviation List

4G	Fourth generation of mobile communication standards
ACR	Absolute Category Rating
ADPCM	Adaptive Differential Pulse Code Modulation
AMR	Adaptive Multi Rate
AMR-WB	AMR wideband
ANOVA	Analysis of Variance
API	Application Programming Interface
ARQ	Automatic Repeat-reQuest
CELP	Code Excited Linear Prediction
CETVSQ	Continuous Evaluation of Time Varying Speech Quality
CI	Confidence Interval
CN	Correspondent Node
CoA	Care-of Address
DCR	Degradation Category Rating
DHCP	Dynamic Host Configuration Protocol
E-model	Synonym for ITU-T Rec. G.107
EDGE	Enhanced Data for GSM Evolution
ETSI	European Telecommunications Standard Institute
FEC	Forward Error Correction
GoP	Group of Pictures
GPRS	General Radio Packet System
GSM	Global System for Mobile Communication
HA	Home Agent
HSDPA	High-Speed Downlink Packet Access
HSPA+	Evolved High Speed Packet Access
HSUPA	High-Speed Uplink Packet Access
IEEE	Institute of Electrical and Electronics Engineers
IETF	Internet Engineering Task Force
iLBC	Internet Low Bitrate Codec
IP	Internet Protocol
ITU	International Telecommunication Union
ITU-T	ITU Telecommunication Standardization Sector
LTE	Long Term Evolution
MIH	Media Independent Handover

MN	Mobile Node
$MOS_{a\|av}$	MOS audio, measured in an audiovisual stimulus
MOS_{av}	MOS audiovisual
MOS_a	MOS audio
$MOS_{int\|av}$	MOS integrated of MOS$MOS_{a\|av}$ and $MOS_{v\|av}$
$MOS_{v\|av}$	MOS video, measured in an audiovisual stimulus
MOS_v	MOS video
MOS_a	Mean Opinion Score
MPEG	Moving Picture Experts Group
MSE	Mean Squared Error
NB	Narrowband
PC	Paired Comparison
PCM	Pulse Code Modulation
PESQ	Synonym for ITU-T Rec. P.862
PLC	Packet Loss Concealment
POLQA	Synonym for ITU-T Rec. P.863
PSNR	Peak Signal to Noise Ratio
QoE	Quality of Experience
QoS	Quality of Service
RMSE	Root Mean Square Error
RTCP	RTP Control Protocol
RTP	Real-time Transport Protocol
SIP	Session Initiation Protocol
SNR	Signal to Noise Ratio
SSCQE	Single Stimulus Continuous Quality Evaluation
SSIM	Structural Similarity Index
SWB	Super wideband
UMTS	Universal Mobile Telecommunications System
VAD	Voice Activity Detection
VoIP	Voice over IP
VQM	Synonym for ITU-T Rec. J.144
VQUAD	Synonym for ITU-T Rec. J.341
WB	Wideband
WB-PESQ	Wideband PESQ, synonym for ITU-T Rec. P.862.2
WiFi	Wireless Fidelity, synonym for WLAN
WLAN	Wireless Local Area Network
XML	Extensible Markup Language

Chapter 1
Introduction

Advances in wireless communications stimulate fast development of multimedia services for mobile networks. Over the top services, such as speech and video telephony, gain the attention of mobile users, who nowadays do not expect connectivity only, but also demand for Quality of Experience (QoE). In this sense, the concept of an "always best connected" user faces new challenges. Nomadic use of services and heterogeneity of technology make it impossible to constantly provide the highest quality, and its active management is required. This tightly couples the success of speech and video telephony offerings in heterogeneous networks with the awareness of user perception of changing quality.

In this sense, this book is tailored to investigate perceptual effects that may occur in heterogeneous wireless networks when speech and video telephony is used. In particular, user perception of quality changes due to switching between networks, codecs, and bit rates during an ongoing transmission is in focus of the analysis. This knowledge is inevitable for a perception-based design of service and mobility management in modern networks.

1.1 Motivation and Objectives

This book was motived by the heterogeneity of technology that can be seamlessly used by nomadic users in modern communication systems [1]. Through the transparent and ubiquitous access to speech and video telephony services, users can move through geographical areas that are covered by diverse wireless networks, while a connection to the service is preserved [2]. However, an ideal communication path does not exist and many fundamental trade-offs must be considered in provisioning the service quality [3]. Network coverage versus throughput and reliability of a connection is one aspect of the decision [4, 5]. Efficiency versus robustness of signal compression is another [6–8]. In this sense, to make the best of available resources when a telephony service is used by nomadic users, active adaptation of ongoing transmission is required.

B. Lewcio, *Management of Speech and Video Telephony Quality in Heterogeneous Wireless Networks*, T-Labs Series in Telecommunication Services,
DOI: 10.1007/978-3-319-02102-7_1,

However, most of the service and mobility management efforts neglect user perception and focus on numeric measurement of system performance instead. Network researchers benchmark the quality of a mobile system using parameters, such as packet loss or throughput [9], as well as connection degradations due to a network handover [10–13]. In this sense, network mobility is managed based on parametric triggers, such as the prediction of user mobility patterns according to Signal-to-Noise Ratio (SNR) changes [14], or to network layer metrics such as variation of inter packet delay [15]. In addition, more advanced efforts towards context-aware mobility management already exist, which combine the knowledge of a mobile station with information from the network to support the decision process [16]. In turn, in high communication layers that reach telephony applications, researchers aim at efficient and network-friendly signal compression [17], which applies modern means of scalable multi-rate coding and of error resiliency [18–20]. These mechanisms are used to optimize the coding process and to balance between compression efficiency and error robustness, either manually, or automatically, according to decoding errors that can be detected [21–23].

Fortunately, in the last decade an increasing number of researchers made the step towards service and mobility management that is based on the knowledge of Quality of Experience. For example, in [24] speech quality is estimated to decide when a handover between WiFi and HSPA networks should be scheduled. In [25], the effect of packet loss in heterogeneous networks is mitigated by a changeover between two narrowband speech codecs, PCMA and GSM, according to parametric quality prediction. In [26–28], the encoding bit rate of speech and video codecs is adjusted to maximize the Quality of Experience of mobile users. However, even though there are many studies that aim at perception-based management of quality in wireless networks, they neglect the fact how the process of quality adaptation, itself, is perceived by the users. Moreover, even though previous studies of user perception of time-varying quality already exist [29–31], this knowledge is not tightly integrated in the management of speech and video telephony in heterogeneous wireless networks. In these networks, however, time-variation is an intrinsic characteristic, and neglecting QoE when the service is dynamically adapted can have tremendous consequences. For example, intended quality improvements, such as switching from narrowband to wideband speech, can actively degrade the overall Quality of Experience, if it is performed too late in a call [32, 33]. Another prominent example is frequent adaptation of transmission that might not improve the overall Quality of Experience [30], and thus can be abandoned, at the same time reducing the effort of service management and improving the scalability of a telecommunication system.

Therefore, it is the goal of this book to analyse the Quality of Experience when speech or video telephony is used in heterogeneous wireless networks. Doing so, user perception of transmission changes, such as roaming between wireless networks, switching the video and speech codecs, and changing the video bit rate during active calls, are in focus of the investigation. These procedures are considered as intrinsic features of modern telecommunication systems, and a set of research questions helps to guide this investigation:

- Which network and codec characteristics are the most relevant ones from the perceptual point of view?
- In which situations the techniques of network, codec, and bit rate switching help to improve the user experience, and when do they degrade the quality perception?
- How are quality changes perceived and what are the most effective strategies for quality adaptation in a call?

Facing the fact that the perception of active quality management in heterogeneous networks is not well understood yet, standardized quality models that are used to predict the perceived quality must be fully validated for the mobile use case, as they have not been designed for this purpose. Even if it is known that the models are not capable to handle certain quality degradations, they are not well linked to underlying effects that are the origin of the problem. Therefore, it is another goal of this research to identify whether the perceptual effects that are quantified in this work can be captured by quality prediction models that are standardized by the Telecommunication Sector of the International Telecommunication Union (ITU-T).

In sum, the research goals that are formed in this section enable to identify the most effective strategies for quality management from a perceptual point of view, as well as to assure valid instrumental methods of quality prediction in heterogeneous networks. Both objectives contribute to maximize the Quality of Experience in modern telephony systems.

1.2 Contribution

In this book, I show that networking research and quality engineering must be complementarily applied in heterogeneous wireless networks to assure high Quality of Experience of speech and video telephony. To implement this, this book includes three main contributions:

- My first contribution is the design and implementation of a research testbed that allows to evaluate user experience of speech and video telephony in heterogeneous wireless networks. The testbed enables to perform passive as well as interactive tests, and to collect data in form of transmitted signals that would be otherwise directly presented to the user, system logs, and network traces. A variety of transmission parameters can be adjusted in real-time, and amongst others, an algorithm that enables to seamlessly switch between various speech and video codecs during an ongoing transmission has been developed. To cope with the multitude of supported features, a testbed controller reduces the complexity of performing an experiment. An event-based concept of test definition allows to automatically generate any amount of data that is required to investigate arbitrary transmission scenarios, including conditions in which the streaming properties are changed multiple times during an ongoing session. On top of the testbed, a demon-

strator is deployed that enables to present the effects that were addressed in this research.
- The next contribution of this book is a study of speech telephony quality in heterogeneous wireless networks. The analysis addresses the user perception of quality, as well as the accuracy of instrumental quality prediction. Results of several laboratory experiments are used to derive guidelines of perception-aware management of quality, as well as to identify the strengths and weaknesses of the existing quality prediction models when they are applied in heterogeneous wireless networks. An additional extensions of the wideband E-model is proposed to enable quality measurement when a switch between diverse speech codecs is expected.
- The last contribution of this book is a study of video telephony quality in heterogeneous wireless networks. This contribution is constructed similarly to the corresponding study of speech quality that was described above, additionally addressing the cross-influence of audio and video quality, as well as validation of existing call quality models in the context of video calls.

1.3 Book Outline

Chapter 2 briefly summarizes background information that is relevant for this book. First of all, the heterogeneity of technology in modern telecommunication systems is discussed, in particular addressing the fundamental trade-offs in networking and in signal compression. Then, the concept of Quality of Experience (QoE) is introduced, and common methods of subjective and instrumental quality assessment are described.

Chapter 3 presents a research testbed that was developed in this work to evaluate user perception of quality in heterogeneous wireless networks. The description of the testbed comprises its network and software architecture, including a description of a system for automatic testbed control, and of a demonstrator that is built on top of the testbed. A particular focus is put on the design, technical realization, and qualitative evaluation of a solution that was developed to switch between speech and video codecs within active calls, which is a core capability of the testbed.

Chapter 4 addresses the Quality of Experience in heterogeneous wireless networks when speech telephony is used. This chapter is divided into three parts. First, the methodology that is applied to conduct the experiments is explained. Then, the results of the conducted user tests are analysed, and some rules of perception-based mobility management are recommended. Finally, based on the results of the user tests, the accuracy of existing quality prediction models in the context of heterogeneous networks is analysed. The strengths and weaknesses of those methods are discussed, and some modelling extensions are proposed.

Chapter 5 addresses the Quality of Experience in heterogeneous networks when video telephony is used. The analysis in this chapter is structured similarly to the analysis of speech telephony in Chap. 4, additionally exploring the interaction between speech and video quality, and validating the existing call quality models in the context of wireless video telephony.

Chapter 6 concludes this book, lists the open issues that arrived out of this work, and discloses plans for further work.

References

1. Huber JF (2004) Mobile next-generation networks. IEEE Multimedia 11(1):72–83
2. Cuevas A, Moreno JI, Vidales P, Einsiedler H (2006) The IMS service platform: a solution for next-generation network operators to be more than bit pipes. IEEE Commun Mag 44(8):75–81
3. Gustafsson E, Jonsson A (2003) Always best connected. IEEE Wirel Commun 10(1):49–55
4. Si P, Ji H, Yu FR (2010) Optimal network selection in heterogeneous wireless multimedia networks. Wireless Netw 16(5):1277–1288
5. Pan YC, Sun YS, Hsu C, Chen MC (2009) A user-decided service model and resource management in a cooperative WiMAX/HSDPA network. In: Proceedings of IEEE international conference on communications (ICC). Dresden, pp 1–6
6. Möller S, Raake A, Kitawaki N, Takahashi A, Wältermann M (2006) Impairment factor framework for wideband speech codecs. IEEE Trans Audio Speech Lang Process 14(6):1969–1976
7. Girod B, Farber N (2000) Wireless video. Marcel Dekker, New York
8. Wiegand T, Schwarz H, Joch A, Kossentini F, Sullivan GJ (2003) Rate-constrained coder control and comparison of video coding standards. IEEE Trans Circ Syst Video Technol 13(7):688–703
9. Stallings W (2000) Data and computer communications, 7th Edn. Prentice Hall, Upper Saddle River
10. Vatn J (1999) Long random wait times for getting a care-of address are a danger to mobile multimedia. In: Proceedings of IEEE international workshop on mobile multimedia communications (MoMuC). San Diego, pp 142–144
11. Pyun Jae-Young (2008) Context-aware streaming video system for vertical handover over wireless overlay network. IEEE Transa Consum Electron 54(1):71–79
12. Pahlavan K, Krishnamurthy P, Hatami A, Ylianttila M, Makela JP, Pichna R, Vallstron J (2000) Handoff in hybrid mobile data networks. IEEE Pers Commun 7(2):34–47
13. Ma L, Yu F, Leung VCM, Randhawa T (2004) A new method to support UMTS/WLAN vertical handover using SCTP. IEEE Wirel Commun 11(4):44–51
14. Chien SF, Liu H, Low ALY, Maciocco C, Ho YL (2008) Smart predictive trigger for effective handover in wireless networks. In: Proceedings of IEEE international conference on communications (ICC). Beijing, pp 2175–2181
15. Cunningham G, Perry P, Murphy L (2004) Soft, vertical handover of streamed video. In: Proceedings of IET international conference on 3G mobile communication technologies. London, pp 432–436
16. Taniuchi K, Ohba Y, Fajardo V, Das S, Tauil M, Cheng YH, Dutta A, Baker D, Yajnik M, Famolari D (2009) IEEE 802.21: media independent handover: features, applicability, and realization. IEEE Commun Mag 47(1):112–120
17. Wenger S (2003) H. 264/AVC over IP. IEEE Trans Circ Syst Video Technol 13(7):645–656
18. Schwarz H, Marpe D, Wiegand T (2007) Overview of the scalable video coding extension of the H. 264/AVC standard. IEEE Trans Circ Syst Video Technol 17(9):1103–1120
19. Bessette B, Salami R, Lefebvre R, Jelinek M, Rotola-Pukkila J, Vainio J, Mikkola H, Jarvinen K (2002) The adaptive multirate wideband speech codec (AMR-WB). IEEE Trans Speech Audio Process 10(8):620–636
20. Wang Y, Zhu QF (1998) Error control and concealment for video communication: a review. Proc IEEE 86(5):974–997

21. Zhang R, Regunathan SL, Rose K (2000) Video coding with optimal inter/intra-mode switching for packet loss resilience. IEEE J Sel Areas Commun 18(6):966–976
22. Girod B, Farber N (1999) Feedback-based error control for mobile video transmission. Proc IEEE 87(10):1707–1723
23. Seo JW, Woo SJ, Bae KS (2001) Study on the application of an AMR speech codec to VoIP. In: Proceedings of IEEE international conference on acoustics, speech and signal processing (ICASSP). Salt Lake, pp 1373–1376
24. Varela M, Laulajainen J (2011) QoE-driven mobility management—integrating the users' quality perception into network-level decision making. In: Proceedings of IEEE international workshop on quality of multimedia experience (QoMEX). Mechelen, pp 19–24
25. Leng Ng S, Hoh S, Singh D (2005) Effectiveness of adaptive codec switching VOIP application over heterogeneous networks. In: Proceedings of international conference on mobile technology, applications, and systems (MobiSys). San Juan, pp 1–7
26. Zhang H, Zhao J, Yang O (2008) Adaptive rate control for VoIP in wireless Ad Hoc networks. In: Proceedings of IEEE international conference on communications (ICC). Beijing, pp 3166–3170
27. Khan A, Sun L, Ifeachor E (2012) QoE prediction model and its application in video quality adaptation over UMTS networks. IEEE Trans Multimedia 14(2):431–442
28. Agboma F, Liotta A (2008) QoE-aware QoS management. In: Proceedings of international conference on advances in mobile computing and multimedia (MoMM). Linz, pp 111–116
29. Voran SD (2005) A basic experiment on time-varying speech quality. Czech Republic. In: Proceedings of international conference on measurement of speech and audio quality in networks (MESAQUIN), Prague
30. Gros L, Chateau N (2001) Instantaneous and overall judgements for time-varying speech quality: assessments and relationships. Acta Acustica United Acustica 87(3):367–377
31. Weiss B, Möller S, Raake A, Berger J, Ullmann R (2009) Modeling call quality for time-varying transmission characteristics using simulated conversational structures. Acta Acustica United Acustica 95(6):1140–1151
32. Möller S, Wältermann M, Lewcio B, Kirschnick N, Vidales P (2009) Speech quality while roaming in next generation networks. In: Proceedings of IEEE international conference on communications (ICC). Dresden, pp 1–5
33. Voran SD (2010) Subjective ratings of instantaneous and gradual transitions from narrowband to wideband active speech. In: Proceedings of IEEE international conference on acoustics, speech and signal processing (ICASSP). Dallas, TX, USA, pp 4674–4677

Chapter 2
Experience of Wireless Telephony

The experience of speech and video telephony in IP-based wireless networks depends on the technology that is used to deliver the service, as well as the quality perception of a user who utilizes it. Accordingly, in this chapter the heterogeneity of technology that is inevitable for packet-based wireless telephony is explained, and the fundamental trade-offs in the domains of networking and signal coding are discussed. Moreover, the concept of Quality of Experience (QoE) is introduced and an overview of methods for subjective and instrumental assessment of perceived quality is provided.

2.1 Technology Heterogeneity

A typical telephony scenario can be schematically represented as an end-to-end communication of two clients that are connected through an IP-based network. To transmit the data, the entities encode the speech and video signal, fragment the encoded bit stream in appropriate chunks that are then encapsulated and sent over the network. The received packets are buffered to restore the correct packet order, decapsulated, and decoded. As a result, a plethora of network and application components interact with each other during a call, in general distorting the quality due to signal coding or transmission errors.

As an ideal communication system does not exist, wireless telephony faces two fundamental trade-offs in the design of a service. A trade-off between throughput, reliability, and coverage of a network is one issue, and a trade-off between efficiency of signal compression and injection of data redundancy that is required to improve the transmission reliability is another [1]. As a matter of fact, active management in both domains is required to maximize the quality of service in a dynamic wireless environment.

B. Lewcio, *Management of Speech and Video Telephony Quality in Heterogeneous Wireless Networks*, T-Labs Series in Telecommunication Services,
DOI: 10.1007/978-3-319-02102-7_2,

2.1.1 Networking Trade-Offs

Packet-based communication poses many challenges in transmission performance and thus a range of parameters was defined by ITU-T to characterize the connection [2]. Metrics such as throughput, packet loss ratio, network delay, and inter packet delay variation (so called "jitter") are commonly used to describe the Quality of Service (QoS) from the network perspective. These parameters are of particular relevance in 4G systems that interconnect a wide variety of wireless networks that differ in the purpose of their design, and so in link-layer characteristics. They were intended to assure ubiquitous computing for various types of services, indoor and outdoor.

In this sense two main groups of wireless networks can be differentiated. Networks that provide a low-bandwidth throughput over a wide geographic area and networks that provide the opposite, a high bandwidth service over a narrow geographic area [3]. Respectively, cellular and wireless local area networks (WLANs) represent both of these groups. The cellular technology was developed to cover wide land areas, which is realized with the concept of adjacent network cells that provide radio access through a base station. The cellular technology evolved over the years, going the way from the first GSM systems through extensions towards packet-based transmission, GPRS and EDGE, upgrades to UMTS, HSDPA, HSUPA, HSPA+, towards LTE [4]. In turn, WLAN networks have been primarily adopted for use in a local context, for example at home, providing flexible access to a reliable wireless connection in a limited range. The WLAN technology also evolved over the years from the first widely accepted IEEE 802.11b standard, through 802.11g, and 802.11n, continuing the efforts towards gigabit WiFi [5]. WLAN capable devices are frequently interchangeably named as WiFi, which is derived from the name of WiFi alliance whose mission is to assure global interoperability of the standard-compliant devices.

Due to the coexistence of different wireless technologies in certain geographical areas an overlaid architecture of wireless networks is created [3]. In this network hierarchy the lowest level is characterized by a small network coverage and high-speed connection, and the highest level represent the opposite, large geographical coverage, and low connection throughput. Having such a system, intelligent management of the connectivity according to user location and utilized service type is required [6]. The attachment point to the Internet must be frequently changed. This process is referred to as a network handover, further differentiating this term in a homogeneous and heterogeneous handover, depending if the handover is performed between access points that belong to the same or to different wireless technologies [7]. In this sense, two types of mobility are introduced. Micro and macro mobility is supported if a terminal changes its attachment point to the Internet within one or across networks managed by different organizations, respectively. If an ongoing connection of a mobile station can be maintained during and after a handover, seamless mobility is assured.

There are several protocols that enable end-devices to change the attachment point to the Internet. Mobile IP [8] or SIP [9] are only two of them that operate in different transmission layers. Both of the approaches introduce certain trade-offs, such as transmission overhead and reliability [10, 11]. Due to the latter feature—higher reliability of the networking-based solution—Mobile IP was applied in this research.

Mobile IPv4 defines mechanisms that enable a mobile terminal to change its point of attachment to the Internet whilst remaining reachable through a permanent address, so called home address, at the same time preserving all the active connections while travelling to a new network. When a mobile node is connected to its home network, no special mode of operation is required, and packets are forwarded using normal IP routing between the mobile node and any other node it is communicating with. When the mobile node roams to a foreign network, the mobile node acquires an IP address belonging to the address space of the new network it is visiting, the Care-of Address (CoA). The new address can be acquired directly via the fundamental DHCP protocol [12], or through the support of an additional Mobile IP entity, a foreign agent. Afterwards, the mobile node announces its CoA by sending a binding update message to a special router in his home network, the home agent. After the new location is registered, the home agent establishes a network tunnel with the remote node in order to redirect any packets addressed to the home address of the mobile station.

However, an intelligent handover between various technologies is still the main challenge of mobility management [13]. Apart from the fact that the continuity of QoS cannot be always assured due to technology heterogeneity [14], also handover-related degradations, such as throughput discontinuity and data loss, are expected [15]. Therefore, considerable work has been done to optimize the mobility management, which includes both, an intelligent prediction when a handover should be scheduled, as well as methods to reduce handover related degradations. Predictive triggers were developed as a function of user mobility patterns that are predicted based on the Signal-to-Noise Ratio (SNR) [16], or other parameters such as jitter [17]. On the other hand, improvements of the mobility protocols to reduce the handover delay and data loss were proposed [7], as well as combined use of mobility protocols is practiced to improve the experience of mobility [18].

A promising concept of context-aware mobility management is developed in the framework of IEEE. The 802.21 draft, Media Independent Handover (MIH), describes a model of cooperation between the network and mobile terminals [19]. Link-layer knowledge is combined with the support of information service deployed in the network to take advanced handover decisions and to determine the optimal roaming direction. In this way, the network-awarness enables to make the best of the trade-off between throughput, reliability, and coverage.

2.1.2 Coding Trade-Offs

An inevitable enabler of wireless IP-based telephony is signal compression that significantly reduces the volume of network traffic. There are several methods of speech and video signal coding, which introduce a fundamental trade-off between efficiency of signal encoding and redundancy improves the error resiliency.

The speech codecs can be classified according to the compression algorithm and to the frequency range that is encoded. So, narrowband (300–3400 Hz), wideband (50–7000 Hz), super wideband (50–14000 Hz), and full-band (20–20000 Hz) speech codecs are available. In this context, use of more advanced coding techniques than Pulse Code Modulation (PCM), such as for example Code Excited Linear Prediction (CELP), enable to significantly reduce the traffic volume, or to considerably improve the signal quality when the transmission bit rate is constrained [20, 21]. For example the reference narrowband codec that is used in fixed-networks, ITU-T Rec. G.711 [22], requires a throughput of 64 kbit/s, while the Adaptive Multi Rate (AMR) codec that is mandatory in UMTS systems offers not much lower speech quality at 12.65 kbit/s [23]. In the wideband domain, ITU-T Rec. G.722, a royalty-free Adaptive Differential Pulse Code Modulation (ADPCM) codec, provides the highest wideband quality during diotic listening. However, it operates at the bit rate of 64 kbit/s that is substantially higher than 23.05 kbit/s that is required by ITU-T Rec. G.722.2 [24], aka "AMR wideband", to provide comparable quality of wideband speech. As a matter of fact, all of the speech codecs cause non-linear degradations [25], and to quantify the corresponding perceptual impairments a test framework has been standardized [26].

Similarly, a variety of codecs is available to compress the video signal. MPEG-2 [27], H.263 [28], MPEG-4 [29], and H.264 [30] are the main video codecs that are available in the modern communication systems. All the codecs are based on similar coding principles, having the block-based hybrid video coding in common [31], but differing in features and enhancements that were developed to improve the coding efficiency [30, 32]. Moreover, the codecs differ in their purpose. The MPEG-2 codec was the key enabler of digital broadcasting, and its successors were tailored for low-bit rate communication, which is of particular relevance for wireless IP-based communication. At the time this document is written, H.264 is probably the most efficient and scalable codec that is standardized and can be used in a wide range of scenarios from low-bit rate streaming to high definition broadcasting.

The quality of encoded video can be expressed by blockiness, blur, and motion jerkiness, which are the three most evident distortion types [33]. Blockiness is visible discontinuity between the neighbouring picture blocks due to block-based picture quantisation. Blur is the result of truncation of high-frequency information, and jerkiness is the effect of frame skipping [34]. The artifacts occur when the system resources are constrained, which is for example the case when the transmission bit rate is constrained or the computation resources are limited.

Because the interconnecting wireless network that is used to transmit the encoded data may introduce transmission errors that create data loss in the network or drop of packets in the application, reconstruction of the original signal might not be possible and decoding artifacts may occur. In case of speech, this may lead to discontinuity and noisiness of the signal [35, 36]. In turn, the visible artifacts in the video domain are additionally defined as slicing and video freeze, both describing loss of parts or of complete video frames, respectively. To cope with the transmission-related signal degradations, there is a number of application mechanisms that can be applied. They are generally grouped in congestion control, error control, and error concealment [37]. All of the approaches essentially differ in the idea of countering the effect of packet loss. While the goal of congestion control is to decrease the transmission bit rate to reduce the network load, error control aims at the opposite, eventually increasing the transmission bit rate to add redundancy and to improve the coding robustness. Error concealment is a decoder based approach to conceal the error by generating artificial signal components, when original information cannot be reconstructed.

The codec based congestion control utilizes the reduction of the coding fidelity that is defined by the rate distortion [38], as well as the reduction of spatial and temporal information. In this sense, encoding scalability provides many parameters that can be adjusted to limit the traffic volume, such as speech coding bandwidth and coding mode, video resolution, frame rate, and quantization parameters. The parameters should be adjusted according to the type of transmitted content, as different perceptual effects are created [39].

In turn, error control is enabled by several techniques that have been proposed in the literature. Three of them, forward error correction (FEC), automatic repeat request (ARQ), and resilient coding are briefly discussed in the following [40]. In general, FEC approaches include redundant information in the encoded bit stream to enable the reconstruction of missing or corrupted data. A prominent example of such an approach is the use of error correction codes to protect the RTP packets [41]. The ARQ approach, in turn, requests for retransmission of data when errors are detected. In fact, ARQ adds additional trade-off dimension in the error control, namely the latency, and even though ARQ proved to be a more efficient technique of error recovery than FEC [1], the additional delay that is required to retransmit a packet makes this technique less suitable for telephony services.

The third discussed technique of error control, error resilient coding, was introduced to reduce the error propagation in the decoded signal. Therefore, the dependency between consecutively encoded audio or video frames is reduced or additional decoding resynchronization points are added to the bit stream. For example, the speech codec iLBC is a block-independent coding approach for error-prone transmission systems at the cost of coding efficiency [42].

In the video domain, the common approach to reduce the error propagation are two types of coding: structured and intra coding. The structured coding is based on picture segmentation in groups that include several macro blocks of a picture, for example in slices. The slices introduce natural resynchronization points for the decoder, and no intra-frame prediction is performed across the boundary of a slice [37]. As a result, this measure limits the visual impact of packet loss, and in addition this method also reduces the probability of a bit error in the network due to decreased packet size, which is of particular relevance in wireless networks [43]. In turn, the intra coding approach limits the temporal propagation of an error through injection of macro blocks, slices, or pictures that are encoded independently in the bit stream [34]. There are also several others more advanced techniques that were developed with the evolution of video coding. Some examples of them are reference picture selection and data partitioning that were introduced in the H.263 standard or flexible macro block ordering, parameters sets, and redundant slices that were developed in the H.264 standard [37]. All of them increase the coding complexity, which as a matter of fact introduces another dimension of the coding trade-off.

The ability of a decoder to conceal transmission errors is the last discussed feature that is essential in error-prone transmission systems. To detect and mitigate a decoding error, techniques were developed that help to bypass any audible decoding gaps by generation of a synthetic signal in the speech domain [44], or that help to reduce visible artifacts through spatial and temporal error recovery in case of video [40]. However, the implementation of packet loss concealment also increases the decoder complexity.

As a result, the above overview showed that an ideal coding configuration that suits all situations in a dynamic wireless environment does not exist. Therefore, active management of signal coding is required to maximize the perceptual quality. Transmission-aware coding mode selection was developed [34, 45], although approaches of active bit rate management in wireless networks are more commonly used [46–48]. Active switching between various speech and video codecs is an additional method that has been (??) pre-published in the course of this work [49, 50].

2.2 Quality of Experience

It has to be considered that telephony services that run on top of heterogeneous communication systems are created to be consumed by human users. Therefore, the acceptance and experience of quality are the key factors of the success of a service. This puts the science on Quality of Experience (QoE) in the central place to offer high-quality telephony. QoE as such was defined in [51] as "the degree of delight or annoyance of the user of an application or service. It results from the fulfillment of his or her expectations with respect to the utility and / or enjoyment of the application or service in the light of the users personality and current state". In this sense, the Quality of Experience is affected by many factors that are related to the way people communicate with each other, the context in which telephony is used, and the service design itself [25]. The last incorporates the communication efficiency that is an intermediate result of transmission quality.

Facing the broad scope of quality aspects that are related to mobile speech and video telephony [52], this section provides an overview of relevant methods for subjective and instrumental quality measurement.

2.2.1 Subjective Measurement

Subjective experiments employ real users that are invited to judge the transmission quality that is provided by a telecommunication system. Although an universal method of quality assessment does not exist, there are several types of subjective experiments that differ in the test protocol and the way the quality is judged [25].

User experiments can be structured into interactive and passive methods. The interactive tests are developed to reproduce a real call situation in a controlled test environment, where two or more test participants interact with each other according to pre-defined scenarios [53, 54]. For example, they discuss postal cards, or act a pre-defined role in a situation that is close to an every-day conversation, such as ordering a pizza or making an appointment with a doctor [55]. Despite the fact that conversational tests are ecologically valid, they place a content-related focus on the users attention. This leads to the effect that users are generally less analytical in their judgements, and detailed information about individual quality features cannot be captured. Therefore, passive tests, such as listening-only or viewing-only experiments, that do not require any interaction between test participants are conducted [56–58]. These procedures are based on the presentation of a pre-recorded signal, which enables the test participants to more analytically judge the quality of the test material, and reduces the complexity of an experiment.

As a compromise, a third form of semi-interactive tests was developed that eliminate the need of interaction between test participants, but enforces an isolated participant to interact with the test system. As a result, a situation that is close to a real call is created. Such a method was developed in [59] and is recommended for call quality measurement in [60]. This procedure describes that five approx. 6–12 s long samples that correspond to utterances of one call participant are presented with pauses in between the stimuli to the participant of the test. In the pauses, the invited person has to fulfill a content related task, such as to answer a question. At the end of such a call emulation, the overall Quality of Experience is judged.

The quality ratings can be collected in various forms, and diverse quality scales are used are used for this purpose [25, 61]. When the overall quality is of interest, Absolute Category Rating (ACR) is recommended. For this purpose, the 5-point MOS scale that quantifies the perceived quality as "excellent" = 5, "good" = 4, "fair" = 3, "poor" = 2, and "bad" = 1 is widely adapted. However, when more subtle quality differences should be grasped, a direct comparison of test stimuli provides more detailed results. When the degradation of a signal in respect to its reference is tested, the Degradation Category Rating (DCR) can be applied. In turn, when two systems under test should be benchmarked, Paired Comparison (PC) is a more suitable approach. In most of the cases a 5-point scale is recommended, but, if higher sensitivity is required, other types of scales are discussed to be more appropriate [61]. Moreover, according to the research context various types of quality information can be collected, such as overall Quality of Experience or intelligibility of a stimulus [25].

The quality scores that are collected during an experiment are commonly averaged across all test participants to obtain the Mean Opinion Score (MOS). In this sense, the MOS score of the audio, video, and audiovisual quality is denoted as MOS_a,[1] MOS_v,

[1] The terms audio and speech quality are used interchangeably in this work, because the scope of this research is limited to speech signals only. Otherwise, the terms must be differentiated.

MOS_{av}. Because the quality ratings depend on the test environment, in particular on the cross-influence of the quality modalities, which is a special case of video telephony, further differentiation of the MOS definition is applied. When the audio and video quality is judged in an audiovisual context, they are referred to as $MOS_{a|av}$ and $MOS_{v|av}$, respectively, and a mathematical combination of both to an integrated audiovisual score is defined as $MOS_{int|av}$.

A separate set of methodologies was developed to explicitly address the quality of time-varying transmission. In this context, the term quality must be further specialised to describe the quality perception in various situations. So, quality that is perceived at any instant of a call is defined as instantaneous quality, and the overall quality impression at the end of a stimulus is expressed as the overall quality [62]. Both types of quality are related through the "recency effect", the way humans remember the quality, which amplifies the importance of instantaneous quality near the time the overall quality is judged [60].

Accordingly, the methods of time-varying quality measurement aim either at continuous rating of quality using a slider, such as the Single Stimulus Continuous Quality Evaluation (SSCQE), or Continuous Evaluation of Time Varying Speech Quality (CETVSQ) evaluation [57, 62], or at sample-based reconstruction of a phone call situation, where the quality of the individual samples that are concatenated to reproduce a phone call situation may vary and is measured in a separate test. The latter method corresponds to the semi-interactive test protocol that was described above.

The methods of time-varying quality measurement were validated by researchers, revealing the basic effects of time-varying quality perception. Degradation of the overall quality due to transmission variation was discovered in [63], and the recency effect was analysed in [63, 64], although the effect was not detected in all experiments [65]. In addition, researchers could prove that quality degradation within a call are perceived quicker than instant quality improvements [63], and that the quality of a call is related to the worst quality periods within the call [66]. Although similar effects are expected in the domain of video telephony, the empirical confirmation is missing. Some of the video effects are confirmed in this work, though.

The research of time-varying quality is also so far not well linked to the technological aspects of heterogeneous wireless networks, and thus the perception of phenomena such as a network handover, codec changeover, or perception of bit rate adaptation is not well understood. In [65], the author presented an experiment that addressed the perception of different patterns of narrowband speech quality in a stimulus that was realized with various narrowband codecs. Moreover, in frame of our research, we have already pre-published results of several subjective experiments. In [67], we empirically analysed user perception of mobility between networks that support narrowband and wideband speech quality, and in [68] we have extended this research towards real conversations. In [69], we addressed user perception of video-only quality in heterogeneous wireless networks, and in [70] user perception of audiovisual content that is typical for video telephony was evaluated. Our work also stimulated further research, in particular the study of switching between narrowband and wideband speech quality was validated in [71], extending the research towards graceful switching between both quality levels.

2.2.2 Instrumental Measurement

Because subjective experiments are a complex way of quality measurement that is cumbersome for real deployments, instrumental methods of quality prediction are developed, which enables to include QoE in the process of management and monitoring of service quality.

A variety of quality prediction models is available [55, 72–74]. They reflect the quality degradation either in terms of technical parameters, such as Mean Squared Error (MSE), Peak Signal to Noise Ratio (PSNR) or Structural Similarity (SSIM) index [75], or attempt to reconstruct user perception using mathematical models. The quality models are used for assessment or planning of telecommunication systems and can be generally subdivided into signal-based and parameter-based approaches.

Signal-based models analyze the physical audio or video signal to estimate the perceived quality. A comparison of the transmitted signal with its reference is one of the methods. The full-reference models in the domain of video quality prediction are represented for example by ITU-T Rec. J.144 that was developed to benchmark coding-related degradations of standard-definition television [76, 77], and by ITU-T Rec. J.341 that was designed to additionally assess transmission related impairments of high-definition television [78]. In the domain of speech quality prediction, ITU-T Rec. P.862 was created to assess the perception of coding and transmission related errors of narrowband telephony, and the model has been extended towards wideband speech in ITU-T Rec. P.862.2 [79, 80]. ITU-T Rec. P.863 was developed to extend the scope of quality estimation to super-wideband speech and to additionally handle new types of degradation such as time warping [81].

The fact that full-reference models require access to both, the reference and the degraded signal, makes the models reliable diagnostic tools, that, however, are not suitable for real-time quality monitoring. For this purpose parameter-based planning models are preferred. These methods utilize numeric metrics, such as packet loss or quantified coding impairment to estimate the Quality of Experience. So, in the domain of speech quality, ITU-T Rec. G.107, the so called E-model, has been standardized to plan and assess the transmission quality, and an extension of this model towards wideband speech is recommended in ITU-T Rec. G.107.1 [82, 83]. In turn, in the domain of video quality prediction, ITU-T Rec. G.1070 was designed to parametrically estimate the quality in packet-based video telephony systems [84].

Excluding the G.1070 standard, all of the above mentioned models are designed to predict the quality of one modality only, either speech or video quality. However, in video telephony systems knowledge of the overall audiovisual quality is required. The integration of audio and video quality into a single audiovisual metric is addressed in ITU-T Rec. J.148 [85] and an integration function is also a part of ITU-T Rec. G.1070.

Due to the fact that all of the above described models were designed for the assessment of short (approx. 4–8 s) signal stimuli, an extra set of models was developed to predict the quality of a stimulus that has the structure of a typical phone call [64, 66]. These approaches estimate the overall call quality by integration of the quality of individual talk spurts; these models have been primary developed for speech telephony, and they are validated for the use case of video telephony in this work.

However, the validation of quality prediction models in heterogeneous networks is still limited. An E-model based approach of modelling time varying speech quality due to bursty packet loss is proposed in [86]. Apart of this, we have pre-published parts of our research work in [87], where we addressed the accuracy of speech quality prediction in heterogeneous mobile networks, in particular addressing the accuracy of signal-based ITU-T P.862.2 in [88]. We have also pre-published an extension of the E-model, that enables to estimate the speech quality when a speech codec is switched in [89]. Finally, in [90] we addressed the accuracy of speech, video, and call quality models in the context of video telephony in heterogeneous networks.

There are also several other non-standardized approaches of quality modeling that are in particular developed for wireless communication. For example, quantification of quality due to obstacles and user motion is shown in [91]. In [92], the authors proposed a model for estimation of audiovisual quality of a video call in wireless networks, in particular relying on the packet error rate and video frame rate for modelling. Furthermore, in [93], a video quality metric that in addition to the above parameters is based on the encoding bit rate and classification of the video content was described. An approach to model different degrees of audio and video channel degradations during wireless transmission was described in [94], and an approach that is based on the relationship between session time, user satisfaction, and quality parameters was introduced in [95]. However, none of them incorporates the fact how users perceive changing quality.

References

1. Girod B, Farber N (2000) Wireless video. Marcel Dekker, New York
2. ITU-T Rec. Y.1540 (2011) Internet protocol data communication service—IP packet transfer and availability performance parameters. In: International telecommunication union, Geneve
3. Stemm M, Katz RH (1998) Vertical handoffs in wireless overlay networks. Mobile Netw Appl 3(4):335–350
4. Dahlman E, Parkvall S, Sköld J, Beming P (2008) 3G Evolution: HSPA and LTE for mobile broadband. Elsevier, Burlington
5. Hiertz G, Denteneer D, Stibor L, Zang Y, Costa XP, Walke B (2010) The IEEE 802.11 universe. IEEE Commun Mag 48(1):62–70
6. Huber JF (2004) Mobile next-generation networks. IEEE Multimedia 11(1):72–83
7. Vidales P (2005) Seamless mobility in 4G systems. Technical report UCAM-CL-TR-656, University of Cambridge, Computer laboratory
8. Perkins C (2002) IP mobility support for IPv4. RFC 3344 (Proposed standard), Internet engineering task force. Updated by RFC 4721
9. Rosenberg J, Schulzrinne H, Camarillo G, Johnston A, Peterson J, Sparks R, Handley M, Schooler E (2002) SIP: Session initiation protocol. RFC 3261 (Proposed standard), Internet engineering task force
10. Kwon TT, Gerla M, Das S (2002) Mobility management for VoIP service: mobile IP vs SIP. IEEE Wirel Commun 9(5):66–75

11. Schulzrinne H, Wedlund E (2000) Application-layer mobility using SIP. Mobile Comput Commun Rev 1(2):47–57
12. Droms R (1997) Dynamic host configuration protocol. RFC 2131 (Draft standard), Internet engineering task force
13. McNair J, Zhu F (2004) Vertical handoffs in fourth-generation multinetwork environments. IEEE Wirel Commun 11(3):8–15
14. Akyildiz IF, Xie J, Mohanty S (2004) A survey of mobility management in next-generation all-IP-based wireless systems. IEEE Wirel Commun 11(4):16–28
15. Pyun Jae-Young (2008) Context-aware streaming video system for vertical handover over wireless overlay network. IEEE Trans Consum Electron 54(1):71–79
16. Chien SF, Liu H, Low ALY, Maciocco C, Ho YL (2008) Smart predictive trigger for effective handover in wireless networks. In: Proceedings of IEEE international conference on communications (ICC). Beijing, pp 2175–2181
17. Cunningham G, Perry P, Murphy L (2004) Soft, vertical handover of streamed video. In: Proceedings of IET international conference on 3G mobile communication technologies. London, pp 432–436
18. Jung J-W, Mudumbai R, Montgomery D, Hyun-Kook K (2003) Performance evaluation of two layered mobility management using mobile IP and session initiation protocol. In: Proceedings of the IEEE global telecommunications conference (GLOBECOM). San Francisco, pp 1190–1194
19. Taniuchi K, Ohba Y, Fajardo V, Das S, Tauil M, Cheng YH, Dutta A, Baker D, Yajnik M, Famolari D (2009) IEEE 802.21: media independent handover: features, applicability, and realization. IEEE Commun Mag 47(1):112–120
20. ITU-T Rec. G.113 (2009) Transmission impairments due to speech processing. International telecommunication union, Geneve
21. ITU-T Rec. G.113-Amendment 1 (2006) New appendix IV provisional planning values for the wideband equipment impairment factor Ie, wb. International telecommunication union, Geneve
22. ITU-T Rec. G.711 (1988) Pulse code modulation (PCM) of voice frequencies. International telecommunication union, Geneve
23. 3GPP TS 26.071 (1999) Mandatory speech CODEC speech processing functions; AMR speech codec; General description. 3GPP
24. ITU-T Rec. G.722.2 (2003) Wideband coding of speech at around 16kbit/s using adaptive multi-rate wideband (AMR-WB). International telecommunication union, Geneve
25. Möller S (2010) Quality engineering: Qaualität kommunikationstechnischer systeme. Springer, Berlin
26. ITU-T Rec. P.833.1 (2008) Methodology for the derivation of equipment impairment factors from subjective listening-only tests for wideband speech codecs. In: International telecommunication union, Geneve
27. Haskell BG, Puri A, Netravali AN (1996) Digital video: an introduction to MPEG-2. Springer, Berlin
28. Cote G, Erol B, Gallant M, Kossentini F (1998) H. 263+: video coding at low bit rates. IEEE Trans Circ Syste Video Technol 8(7):849–866
29. Li W (2001) Overview of fine granularity scalability in MPEG-4 video standard. IEEE Trans Circ Syste Video Technol 11(3):301–317
30. Wiegand T, Sullivan GJ, Bjontegaard G, Luthra A (2003) Overview of the H. 264/AVC video coding standard. IEEE Trans Circ Syst Video Technol 13(7):560–576
31. Wiegand T, Zhang X, Girod B (1997) Block-based hybrid video coding using motion-compensated long-term memory prediction. In: Proceedings of picture coding symposium. Berlin, pp 1–6
32. Wiegand T, Schwarz H, Joch A, Kossentini F, Sullivan GJ (2003) Rate-constrained coder control and comparison of video coding standards. IEEE Trans Circ Syst Video Technol 13(7):688–703
33. Yuen M (2005) Coding artifacts and visual distortions. In: Wu HR, Rao KR (eds)Digital video image quality and perceptual coding, pp 87–122

34. Zhang R, Regunathan SL, Rose K (2000) Video coding with optimal inter/intra-mode switching for packet loss resilience. IEEE J Sel Areas Commun 18(6):966–976
35. Wältermann M (2013) Dimension-based quality modeling of transmitted speech. Springer, Berlin
36. Côté N (2010) Integral and diagnostic intrusive prediction of speech quality. Springer, Berlin
37. Wenger S (2003) H. 264/AVC over IP. IEEE Trans Circ Syst Video Technol 13(7):645–656
38. Sullivan GJ, Wiegand T (1998) Rate-distortion optimization for video compression. IEEE Signal Process Mag 15(6):74–90
39. Belmudez B, Möller S (2011) An approach for modeling the effects of video resolution and size on the perceived visual quality. In: Proceedings of IEEE workshop on multimedia quality of experience (MQoE). Dana Point, pp 464–469
40. Wang Y, Zhu QF (1998) Error control and concealment for video communication: a review. Proc IEEE 86(5):974–997
41. Li A (2007) RTP payload format for generic forward error correction. RFC 5109 (Proposed standard). Internet engineering task force
42. Andersen S, Duric A, Astrom H, Hagen R, Kleijn W, Linden J (2004) Internet low bit rate codec (iLBC). RFC 3951 (Experimental). Internet engineering task force
43. Stockhammer T, Hannuksela MM, Wiegand T (2003) H. 264/AVC in wireless environments. IEEE Trans Circ Syst Video Technol 13(7):657–673
44. Wah BW, Su X, Lin D (2000) A survey of error-concealment schemes for real-time audio and video transmissions over the internet. In: Proceedings of IEEE international symposium on multimedia software engineering. Taipei, pp 17–24
45. Girod B, Farber N (1999) Feedback-based error control for mobile video transmission. Proc IEEE 87(10):1707–1723
46. Zhang H, Zhao J, Yang O (2008) Adaptive rate control for VoIP in wireless ad hoc networks. In: Proceedings of IEEE international conference on communications (ICC). Beijing, China, pp 3166–3170
47. Khan A, Sun L, Ifeachor E (2012) QoE prediction model and its application in video quality adaptation over UMTS networks. IEEE Trans Multimedia 14(2):431–442
48. Agboma F, Liotta A (2008) QoE-aware QoS management. In: Proceedings of international conference on advances in mobile computing and multimedia (MoMM). Linz, pp 111–116
49. Lewcio B, Möller S, Vidales P (2011) A method for seamless Codec changeover during active video calls. In: Proceedings of IEEE computer-aided modeling analysis and design of communication links and network (CAMAD). Kyoto, pp 133–137
50. Wältermann M, Lewcio B, Vidales P, Möller S (2008) A technique for seamless VoIP-Codec switching in next generation networks. In: Proceedings of IEEE international conference on communications (ICC). Beijing, pp 1772–1776
51. Le Callet P, Möller S, Perkis A (eds) (2012) Qualinet white paper on definitions of quality of experience. In: European network on quality of experience in multimedia systems and services (COST Action IC 1003), Lausanne
52. Jumisko-Pyykkö S, Häkkinen J, Nyman G (2007) Experienced quality factors-qualitative evaluation approach to audiovisual quality. In: Proceedings of IS&T SPIE 19th annual symposium of electronic imaging. San Jose
53. ITU-T Rec. P.805 (2007) Subjective evaluation of conversational quality. International telecommunication union, Geneve
54. Belmudez B, Möller S (2011) Description of conversation scenarios for interactive audiovisual communication. ITU-T Contr. COM 12–271, Geneve
55. Möller S (2000) Assessment and prediction of speech quality in telecommunications. Kluwer Academic Publishers, Boston
56. ITU-T Rec. P.800 (1996) Methods of subjective determination of transmission quality. International telecommunication union, Geneve
57. ITU-T Rec. P.910 (1998) Subjective video quality assessment methods for multimedia applications. International telecommunication union, Geneve

58. ITU-T Rec. P.911 (1998) Subjective audiovisual quality assessment methods for multimedia applications. International telecommunication union, Geneve
59. Berger J, Hellenbart A, Weiss B, Möller S, Gustafsson J, Heikkila G (2008) Estimation of quality per call in modelled telephone conversations. In: Proceedings of IEEE international conference on acoustics, speech and signal processing (ICASSP). Las Vegas, pp 4809–4812
60. ETSI 102 506 v. 1.2.1 (2006) Speech processing, transmission and quality aspects (STQ); estimating speech quality per call. In: European telecommunications standard Institute, Sophia Antipolis
61. S. Möller (2009) Towards a universal scale for perceptual value. ITU-T Contr. COM 12–81, Geneve
62. ITU-T Rec. P.880 (2004) Continuous evaluation of time-varying speech quality. In: International telecommunication union, Geneve
63. Gros L, Chateau N (2001) Instantaneous and overall judgements for time-varying speech quality: assessments and relationships. Acta Acustica United Acustica 87(3):367–377
64. Rosenbluth J (1998) Testing the quality of connections having time varying impairments. ITU-T Contr. D.64, Geneve
65. Voran SD (2005) A basic experiment on time-varying speech quality. Czech Republic. In: Proceedings of international conference on measurement of speech and audio quality in networks (MESAQUIN), Prague
66. Weiss B, Möller S, Raake A, Berger J, Ullmann R (2009) Modeling call quality for time-varying transmission characteristics using simulated conversational structures. Acta Acustica United Acustica 95(6):1140–1151
67. Möller S, Wältermann M, Lewcio B, Kirschnick N, Vidales P (2009) Speech quality while roaming in next generation networks. In: Proceedings of IEEE international conference on communications (ICC). Dresden, pp 1–5
68. Möller S, Wältermann M, Lewcio B, Kirschnick N, Vidales P (2008) Handovers between narrowband and wideband speech transmission. In: Proceedings of sprachkommunikation 2008, Vorträge der ITG-Fachtagung, ITG-Fachbericht. Aachen, VDE-Verlag
69. Lewcio B, Belmudez B, Mehmood A, Wältermann M, Möller S (2011) Video quality in next generation mobile networks—perception of time-varying transmission. In: Proceedings of IEEE international workshop on communications quality and reliability (CQR). Naples, pp 1–6
70. Lewcio B, Belmudez B, Enghardt T, Möller S (2011) On the way to high-quality video calls in future mobile networks. In: Proceedings of IEEE international workshop on quality of multimedia experience (QoMEX). Mechelen, pp 43–48
71. Voran SD (2010) Subjective ratings of instantaneous and gradual transitions from narrowband to wideband active speech. In: Proceedings of IEEE international conference on acoustics, speech and signal processing (ICASSP). Dallas, TX, USA, pp 4674–4677
72. Raake (2006) Speech quality of VoIP: assessment and prediction. Wiley, New York
73. Winkler S, Mohandas P (2008) The evolution of video quality measurement: from PSNR to hybrid metrics. IEEE Trans Broadcast 54(3):660–668
74. Takahashi A, Hands D, Barriac V (2008) Standardization activities in the ITU for a QoE assessment of IPTV. IEEE Commun Mag 46(2):78–84
75. Wang Z, Lu L, Bovik AC (2004) Video quality assessment based on structural distortion measurement. Signal Process Image Communn 19(2):121–132
76. ITU-T Rec. J.144 (2004) Objective perceptual video quality measurement techniques for digital cable television in the presence of a fullrReference. In: International telecommunication union, Geneve
77. Pinson MH, Wolf S (2004) A new standardized method for objectively measuring video quality. IEEE Trans Broadcast 50(3):312–322
78. ITU-T Rec. J.341 (2011) Objective perceptual multimedia video quality measurement of HDTV for digital cable television in the presence of a full reference. In: International telecommunication union, Geneve

79. ITU-T Rec. P.862 (2001) Perceptual evaluation of speech quality (PESQ): an objective method for end-to-end speech quality assessment of narrow-band telephone networks and speech codecs. In: International telecommunication union, Geneve
80. ITU-T Rec. P.862.2 (2005) Wideband extension to recommendation P.862 for the assessment of wideband telephone networks and speech codecs. In: International telecommunication union, Geneve
81. ITU-T Rec. P.863 (2011) Perceptual objective listening quality assessment. In: International telecommunication union, Geneve
82. ITU-T Rec. G.107 (2011) The E-model, a computational model for use in transmission planning. In: International telecommunication union, Geneve
83. ITU-T Rec. G.107.1 (2011) Wideband E-model. In: International telecommunication union, Geneve
84. Rec ITU-T, G (2012) (1070) Opinion model for video-telephony applications. In: International telecommunication union, Geneve
85. ITU-T Rec. J.148 (2003) Requirements for an objective perceptual multimedia quality model. In: International telecommunication union, Geneve
86. Clark A (2001) Modeling the effects of burst packet loss and recency on subjective voice quality. In: Proceedings of IP telephony workshop. New York, pp 123–127
87. Lewcio B, Wältermann M, Vidales P, Raake A, Möller S (2009) Performance of instrumental speech quality measures for next generation wireless networks. In: Proceedings of NAG/DAGA. Rotterdam, pp 1–4
88. Mehmood MA, Lewcio B, Vidales P, Feldmann A, Möller S (2010) Understanding signal-based speech quality prediction in future mobile communications. In: Proceedings of IEEE international conference on communications (ICC). Capetown, pp 1–6
89. Lewcio B, Wältermann M, Möller S, Vidales P (2009) E-model supported switching between narrowband and wideband speech quality. In: Proceedings of IEEE international workshop on quality of multimedia experience (QoMEX). San Diego, pp 98–103
90. Lewcio B, Möller S (2012) Predicting video call quality in future mobile networks. In: Proceedings of IEEE international workshop on quality of multimedia experience (QoMEX). Yarra Valley, pp 140–145
91. Ennaji Y, Boulmalf M, Alaoui C (2009) Experimental analysis of video performance over wireless local area networks. In: Proceedings of international conference on multimedia computing and systems (ICMCS). Chennai, pp 488–494
92. Goudarzi M, Sun L, Ifeachor E (2010) Audiovisual quality estimation for video calls in wireless applications. In: Proceedings of IEEE global telecommunications conference (GLOBECOM). Miami, pp 1–5
93. Khan A, Sun L, Ifeachor E (2009) Content clustering based video quality prediction model for MPEG4 video streaming over wireless networks. In: Proceedings of IEEE international conference on communications (ICC). Dresden, pp 1–5
94. Belmudez B, Möller S, Lewcio B, Raake A, Mehmood A (2009) Audio and video channel impact on perceived audio-visual quality in different interactive contexts. In: Proceedings of IEEE international workshop on multimedia signal processing (MMSP). Rio de Janeiro, pp 1–5
95. Huang TY, Kuan-Ta C, Huang P, Chin-Laung LEI (2008) A generalizable methodology for quantifying user satisfaction. IEICE Trans Commun 91(5):1260–1268

Chapter 3
Heterogeneous Testbed

An analysis of user perception of quality in heterogeneous wireless networks requires a reliable test environment that enables to emulate relevant transmission phenomena under controlled conditions [1–4]. A testbed that enables to link the user experience of speech and video quality to cross-layer transmission parameters in heterogeneous wireless networks is presented in this chapter [5]. The testbed consists of a network infrastructure, of a speech and video telephony client, and of a control module. A large variety of transmission parameters can be customized in the setup, and test data in form of system logs and processed signals can be recorded. Speech, audio, and video codecs, their transmission rate, as well as the size of transmitted packets or the properties of the jitter buffer can be configured. Simultaneously, the network connection can be selected from heterogeneous technologies, and the transmission quality can be artificially degraded. A particular speciality of the testbed is its ability to adjust the parameters during an ongoing transmission, which is essential for a study of time-varying quality. Amongst others, an algorithm that enables to switch between various speech and video codecs during an ongoing transmission has been developed in scope of this work. Despite the multitude of supported features, the testbed was designed to reduce the complexity of performing a test. Through an event-based concept of test definitions that was also developed during this work, test conditions can be automatically processed in the setup. Finally, the testbed can also be used as a demonstration environment.

3.1 Network Architecture

The networking infrastructure of the testing environment consists of two interconnected testbeds: the Mobisense and the Router Lab testbed [6, 7]. While the Router Lab enables to emulate diverse characteristics that are typical for backbone networks, the Mobisense setup provides access to heterogeneous wireless technologies, such

B. Lewcio, *Management of Speech and Video Telephony Quality in Heterogeneous Wireless Networks*, T-Labs Series in Telecommunication Services,
DOI: 10.1007/978-3-319-02102-7_3, © Springer International Publishing Switzerland 2014

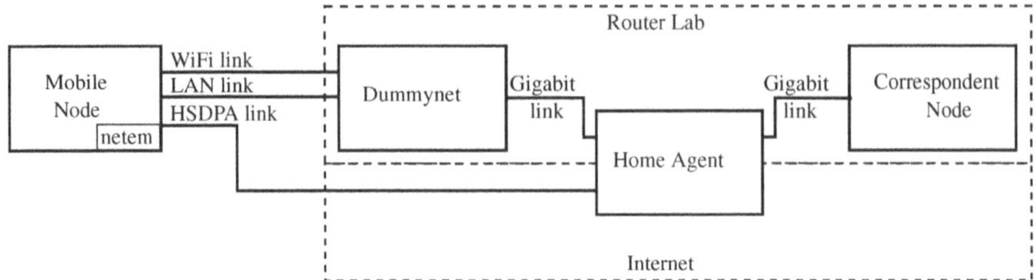

Fig. 3.1 Networking infrastructure of the testbed

as WiFi or UMTS/HSDPA. All of the networks are IP-based and the Mobile IP protocol [8] is used to support the mobility.

The networking architecture of the testbed (cf. Fig. 3.1) consists of three main components: a Mobile Node (MN), a Correspondent Node (CN), and a Home Agent (HA). The Mobile Node is connected via a WiFi and a LAN link to the Router Lab, and via a public HSDPA network to the Internet. The Correspondent Node is connected via a Gigabit LAN link to the home network. The Correspondent and the Mobile Node communicate with each other via a dual-homed Home Agent in both directions, and selectively over the Internet or through the Router Lab. The Home Agent is dual-homed, i.e. connected to both the Internet and the Router Lab, because the access to 3G technologies is available through a public UMTS/HSDPA network only. Therefore, a guarantee of resource allocation was not possible, which was in particular critical for the bit rate consuming video traffic. Therefore, to control the connection quality, traffic from the Internet is routed through the netem module that is installed on the Mobile Node, and each connection that is configured in the Router Lab is routed through the Dummynet node.

The configuration of the main nodes of the testbed is summarized in Table 3.1. The Home Agent is deployed on a Cisco 7204 VXR router with Cisco IOS 12.1, and the Correspondent and the Mobile Node are installed on Dell Optiplex 780 desktop computers that are equipped with Intel(R) Core(TM)2 Duo E7500 @ 2.93 GHz processors and Debian Linux 2.6.32 and Ubuntu Linux 2.6.28-19 as operating systems of the Correspondent and the Mobile Node, respectively. The desktop computers are selected for the setup to avoid any performance issues that may occur during simultaneous speech and video coding, packet streaming, and data recording. All of the processes have to run simultaneously when an experiment that addresses real-

Table 3.1 Configuration of the main testbed nodes

	MN	CN	HA
Hardware	Dell OptiPlex 780	Dell OptiPlex 780	Cisco 7204 VXR
OS	Ubuntu Linux 2.6.28-19	Debian Linux 2.6.32	Cisco IOS 12.1
Software	netem, nattt, lmipd	-	-

time communication is performed. Additionally, the Mobile Node was extended with additional software modules as follows. A *lmipd* client was installed to support Mobile IP, a *nattt* module is used to enable mobility behind NAT, and a netem module [9] was enabled to artificially degrade the quality of the HSDPA connection. On both, the Mobile and the Correspondent Node, *tcpdump* was installed to collect network traces on all relevant network interfaces.

An additional emphasis is put on the approach of mobility control that is applied in the testbed. The network handover is initiated manually, which means that no automatic triggers from the network are used for this purpose. Furthermore, when a handover is performed, the "make-before-break" policy is applied. This means that both of the involved network interfaces are active during a handover and the network can be received via both of them until the procedure is completed. This approach is different from the "break-before-make" policy where the mobile station starts to sense for a new network after the original connection has been interrupted. Moreover, the physical location of the Mobile Node is always fixed, which assures constant strength of the wireless signal, and any link-layer degradations that occur in the setup are generated using the respective tools. This approach is selected, because amongst others this work was focused on user perception of a handover between heterogeneous networks, rather than instability of the wireless signal.

As a result, the assembled setup created an experimental platform for evaluation of user perception of mobility, where apart from the network handover between heterogeneous technologies, diverse networking conditions such as degradations due to packet loss or variable network delay, but also degradations due to background traffic can be investigated.

3.2 Speech and Video Telephony Client

To enable speech and video calls between the Mobile and the Correspondent Node, a test client was deployed on the computers. The test application was developed based on PJPROJECT 0.8.3 that is a VoIP framework [10] in which the Session Initiation Protocol (SIP) and Real-time Transmission Protocol (RTP) [11, 12] are used to control multimedia sessions and to stream real-time data, respectively. The application from [10] has been extended in several ways during this work. The video support has been implemented, and an additional set of features, including a technique for switching between active codecs (cf. Sect. 3.3), has been added to enable research on time-varying quality in heterogeneous networks. The respective research capabilities are explained in this section along with the general architecture of the application.

The software architecture of the test client (cf. Fig. 3.2) consists of two media flows of similar structure that are used to process the speech and video signal. While the speech part of the framework was developed by the authors of PJPROJECT, the video support has been implemented in this work through integration of libavformat, libswscale, and libavcodec libraries [13]. Each media flow includes a signal

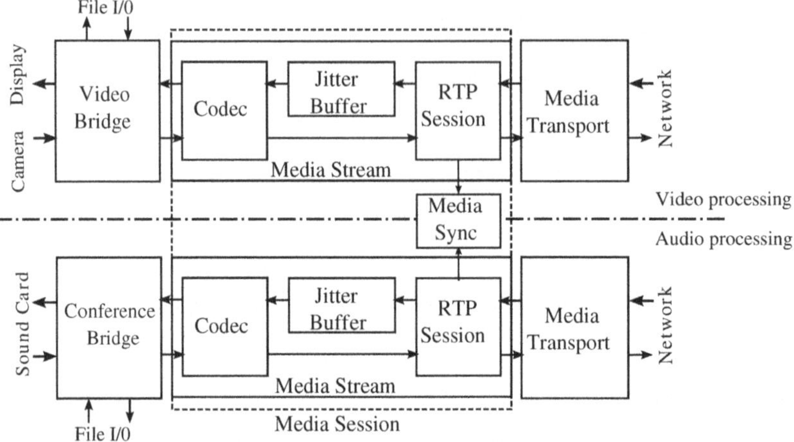

Fig. 3.2 Software architecture and media flow in the test client

bridge, either a Conference or a Video Bridge, that interconnect sources and sinks of uncompressed signals, such as peripheral devices, File I/O, and Media Streams.

A Media Stream is used to transmit multimedia data through the network. It consists of a Codec, a Jitter Buffer, and a RTP Session unit. Each Media Stream is on one end bidirectionally attached to a signal bridge and on the other end it is bidirectionally attached to a Media Transport that controls network sockets of the RTP/RTCP protocol. Multiple Media Streams, for example one audio and one video stream, can be combined to a Media Session. In such a case, the streams are additionally attached to a Media Sync that tracks the synchronization between the streams, which is based on the timestamp information that is provided by the RTP/RTCP protocol.

Accordingly, packets that arrive from the network are forwarded by the Media Transport to the RTP session, where the RTP header is validated. Correctly received payloads are decapsulated and stored in the Jitter Buffer. From there, data is fetched by the Codec, and the decoded speech or video signal is passed to a signal bridge. The bridge forwards the decoded signal to a display, sound card, or to a file recorder. In turn, to send speech or video data, uncompressed signals from a camera, sound card, or from a file reader are passed to the Codec that encodes and, if necessary, fragments the stream to create appropriate payload chunks. The payloads are then forwarded to the RTP Session where a RTP header is attached and thereafter the packets are sent via the Media Transport to the network. The packetization of the audio and video streams follows RFC standards for single channel communication, and both of the streams are transmitted separately; the audio and video packets are not multiplexed.

The modularity of the above described software architecture enabled to extend the application in several ways. The resulting research features are described in groups according to the functional layer in the application they apply to. In the following, the processing of uncompressed signals are introduced first. Next, the reconfigurability of the speech and video coding is presented. Last, the features that are related to packetization and buffering are described.

3.2.1 Raw Signal Processing

As already mentioned in the general overview, the raw signal in the application is routed through a signal bridge to several modules. Many of them are used as a signal output. For example, the audio signal can be played-out through the sound card, and the video signal can be displayed through a number of graphical user interfaces (GUIs). Any GUI that is developed as a shared-object library and that matches the API of the test client can be dynamically loaded, which is a feature of the application that enables to flexibly test various user interfaces. Currently, there are three GUIs that are implemented in the setup:

- A demonstration GUI that is described in Sect. 3.5.
- A stand-alone video display that is used to monitor the video signal in the application.
- An interface for video telephony tests.

Moreover, the uncompressed signal can also be recorded to disk, which enables to capture signal degradations that are directly perceived by the users. The set of degradations that are related to real-time communication includes for example artifacts that occur due to data loss in the jitter buffer, or video freezing that occurs due to buffer underflows. These types of degradations cannot be captured when the encoded data or network traces are captured.

In turn, the input modules that are integrated in the test application enable to feed the system with uncompressed speech and video signals that are stored in files or that are acquired from peripheral devices. The support of file reading is a key enabler of experiments that require the use of pre-recorded material. Signal acquisition from peripheral devices, such as sound cards or cameras, is a key enabler of interactive tests. In the current implementation of the test client two types of camera drivers, firewire and "video for linux"(v4l), are integrated. The firewire implementation is used in laboratory tests to achieve high quality of the video signal, and the v4l support is used to attach build-in laptop cameras and USB devices that are integrated in common devices.

Table 3.2 Raw signal formats and video scalers that are supported by the test client

Parameter	Supported options
Audio clock rate (kHz)	8, 16, 32, 44.1, 48
Video format	yuv420p, yuv422, rgb24, yuv422p, yuv444p, yuv410p, yuv411p, pal8, yuvj420p, yuvj422p, yuvj444p, yuvj440p, uyvy422
Video scaler	Fast bilinear, bilinear, bicubic, x, point, area, bicublin, gauss, sinc, Lanczos, spline

When a format mismatch between the interconnected signal modules is detected, both, the Conference and the Video Bridge, support certain features to transform the signal and assure the interoperability. The Conference Bridge is capable of re-sampling the audio signal, and the Video Bridge is enabled to convert the format, to rescale, and to deinterlace the video frames, if necessary (cf. Table 3.2 for the list of the supported video formats and rescaling techniques). Through this design, quality experiments can be arranged by interconnecting modules with miss-matching characteristics. For example the effect of rescaling can be analysed by attaching a full screen video display to a Media Stream that operates at a lower resolution. Moreover, the number of video frames that are processed per second in the Video Bridge can be arbitrarily configured. However, it has to be emphasised that only video modules that support the configured frame rate can be interconnected. This imposes a flexibility issue, and to enable the signal routing between modules of different frame rate, a conversion technique must be implemented.

The presented concept of the signal routing in the test application enables to use the test client for various types of tests. Interactive and non-interactive tests can be conducted and various peripherial devices and user interfaces can be used. Despite the fact that the test material can also be generated offline, any effects that would be perceived by the users, such as video freezing, decoding or rescaling artifacts can be recorded. This empowers the testbed to capture any artifacts that may occur in the entire communication chain of a real-time communication system.

3.2.2 Speech and Video Coding

In any real-time communication system, so in the test application, the signal is compressed before the transmission through the network. To enable research on quality impairments due to signal coding in the presented testbed, a set of codecs that are typical for real-time communication has been integrated in the test client during this work. In case of audio and speech, reference implementations that are provided by ITU-T, ETSI, or 3GPP were used. In case of video, codec implementations that are included in the *libavcodec* library were integrated (cf. Table 3.3). All of the codecs can be flexibly used in the test environment.

Table 3.3 Speech, audio, and video codecs supported in the test client

Type	Name
Speech narrowband	G.711, G.723.1, G.726, AMR, GSM, Speex
Speech wideband	G.722, G722.2, G.711.1, Speex
Speech super-wideband	G.722.1C
Audio fullband	G.719, CELT
Video	MPEG-2, MPEG-4, H.263+, H.264

The speech and audio codecs can be selected from a pool of narrowband (300–3400 Hz), wideband (70–7000 Hz), super-wideband (50–14000 Hz), and full band (20–20000 Hz) codecs whose input is filtered with a suitable band pass filter that is recommended in ITU-T Rec. G.191. Moreover, the encoding bit rate of the codec can be adjusted, if this feature is foreseen in the codec design. In case of the video codecs, additionally the resolution of the input and the Group of Pictures (GoP) structure can be modified.

The codecs in both, the audio and the video stream, can be switched during an ongoing session, which is an essential capability of the presented testbed. For this purpose, an algorithm for seamless speech and video codec changeover has been developed (cf. Sect. 3.3). The solution is based on a SIP handshake that is used to initiate the codec changeover procedure and on a graceful replacement of Media Streams during ongoing transmission. The codec changeover as such was evaluated in this work as a technique for transmission adaptation in heterogeneous wireless networks.

As a result, through the multitude of codecs that are integrated in the testbed, and through the reconfigurability of the coding parameters, diverse audio and video coding configurations can be evaluated. As a consequence, the perceptual trade-offs between audio and video quality can be explored, and, in particular, strategies for adaptation of the perceived quality during an ongoing transmission can be developed.

3.2.3 Packetization and Buffering

Another key research feature of the presented test client is the reconfigurability and monitoring of the jitter buffer parameters, as well as the reconfigurability of the maximal size of the video packets that are transmitted through the network. Any compressed payload of either speech or video signal has to be encapsulated in an RTP packet before the transmission, and a set of IETF recommendations, such as [14, 15], is provided to the implementors for this purpose. A particular use case of this process is related to the video packetization, where the size of a compressed video frame may frequently change and in many cases it exceeds the size of the Maximal Transfer Unit (MTU) of the underlying network. Those frames must be fragmented prior to the encapsulation. Therefore, two configuration approaches have been integrated in the client during this work, to restrict the size of the generated payloads. The number of codec syntax units, such as MPEG slices, or H.264 Network Abstraction Layer (NAL) units that are included in a packet can be configured, or the byte-size of the payloads can be explicitly limited.

In the context of payload fragmentation, the design of the jitter buffer that is used to ensure continuous playback of the received data in real time communication systems plays a major role. To fulfill this task the jitter buffer that is provided in the PJPROJECT framework makes use of the principle of circular buffering [16]. This means that multimedia data in the buffer is forwarded to the decoder by successive shifting of the writing and reading pointers in a circular queue whose size can be dynamically

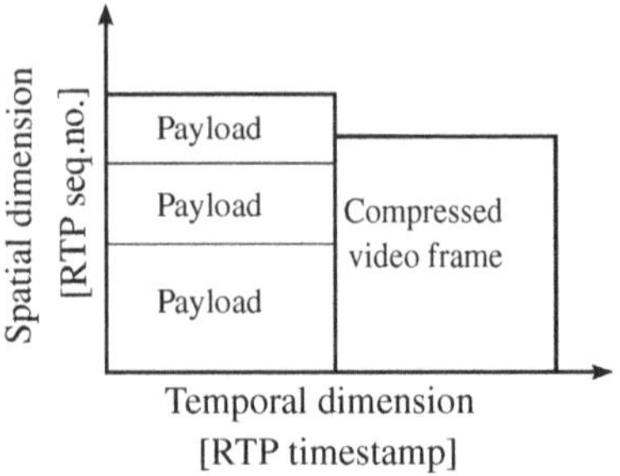

Fig. 3.3 Conceptual view of two-dimensional buffer management

adjusted. Accordingly, the arriving RTP packets are stored in pre-allocated slots of the buffer according to the timestamp information that is included in the RTP header.

However, the original design of the jitter buffer in PJPROJECT does not allow to forward video frames of variable size, and thus a new buffering model has been developed during this work. The direct relationship between the timeline of the buffer and the arrival of the individual packets that convey fragmented payload was detached, and a two-dimensional scheme of the buffer management has been implemented (cf. Fig. 3.3). The new buffering dimensions included:

- A temporal dimension, where the jitter buffer is organized video-frame-wise and is managed based on the RTP timestamps.
- A spatial dimension, where each video frame slot is organized payload-wise and is managed based on RTP sequence numbers.

Both of the buffer dimensions in the application can be configured, and the minimal and maximal size of both, as well as the initial pre-buffering time before the data forwarding can be started. If the adjustment is incorrect and if the configuration does not match the video stream properties and network circumstances, the arriving packets are dropped in the application. This type of data loss is typical for real-time communication systems, but it cannot be measured with classical network monitoring tools. To overcome this limitation and extend the research capabilities of the testbed, an additional jitter buffer monitoring system has been integrated in the test client. As a result, the following statistics of the buffer can be collected during transmission:

- Average, minimal and maximal size of the buffer (in miliseconds).
- Average, minimal and maximal size of the received video frames (in Bytes).
- Statistics about lost and dropped packets and thereof resulting number of corrupted video frames.
- List of missing RTP packets and list of corrupted video frames.
- Overall buffer throughput in Bytes and throughput bit rate.
- Number of buffer underflows and overflows.

The reconfigurability of the packetization and buffering algorithms described in this section, as well as the additional system for monitoring of the jitter buffer enable to perform research experiments on optimization of real-time streaming in heterogeneous wireless networks.

3.3 Codec Changeover Technique

3.3.1 Requirement Analysis

A successful codec changeover during an ongoing real-time communication is performed in such a manner that neither the session nor the continuity of the signal are interrupted. This aim can be achieved in two steps. First of all, the involved communication endpoints have to re-negotiate the coding scheme during an ongoing session and agree on a new codec that is supported by both of the nodes. As a next step, the codecs must be replaced in such a way that any decoding artifacts that might be perceived by the user are reduced as far as possible.

The negotiation of a codec that is used in a SIP-based communication system is standardized in [17]. The procedure relies on a three-way SIP handshake and applies the Session Description Protocol (SDP) to exchange a detailed description of a multimedia session. An example of such a handshake is depicted in Fig. 3.4. The handshake is initiated by an INVITE message that conveys the request for new session parameters. For example, the request for a codec changeover may by signaled as a reordered list of codecs in the attached SDP description. Codecs located on top of this list have the highest selection priority. If the responder supports the proposed configuration, a 200 OK message is issued. The handshake is finished with a final acknowledgment (ACK) that is sent by the requester.

After the codec negotiation is completed, both endpoints should be able to switch to the new encoding scheme. To reduce any signal artifacts, such as decoding gaps that might occur during this operation, several aspects of the application design must be considered. An element of particular relevance for a successful codec changeover is the jitter buffer. As already mentioned in the previous section, this entity is used in a communication system to compensate for a variable delay of arriving RTP packets and to ensure continuous data playback. For this purpose, the arriving packets are stored in the buffer prior to the play-out, and packets that arrive too soon or too late are dropped along the way. Therefore, the primary prerequisite of a seamless codec changeover is no data loss in neither the network nor in the jitter buffer. As a result Eq. 3.1 must apply for every codec that is involved in a changeover.

$$p_r = (1 - p_l) \times P[n_j < D] \tag{3.1}$$

The probability p_r that a packet is timely decoded depends on two factors: the probability p_l that a packet j is lost in the network during the transmission, and the probability that a packet is timely received, which is expressed by the delay probability distribution P with the network delay n_j as a random variable, and the play-out delay of the buffer D.

If the primary prerequisite is fulfilled, $p_r = 1$, and the packet stream is not affected by any data loss, the following concerns about the application design must be taken:

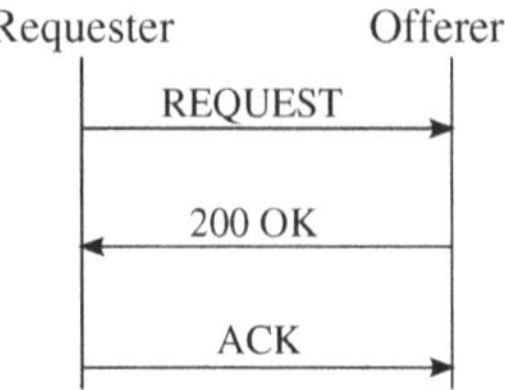

Fig. 3.4 SIP handshake

- Loss of any data that is already stored in the jitter buffers must be prevented. Otherwise, the play-out of the streams must be synchronized.
- Pre-buffering time that a jitter buffer requires before data can be forwarded to the decoder must be assured.
- Initial coding delays, that are in particular observed in case of video codecs, must be bypassed.
- Packets that are received from both codecs may not be rejected on the interface during a changeover.
- Data should not be streamed in a duplicated manner to avoid an increased bandwidth utilization.

Only if the above issues of application design are considered and the primary prerequisite is fulfilled a seamless codec changeover is possible.

3.3.2 Implementation

The requirements that were presented in the previous subsection were considered during the development of a codec changeover technique for the test application (cf. Fig. 3.5). To prepare a changeover, a second Media Stream that will handle the new codec is created and is configured to the same state as the initial stream. This means that the RTP statistics as well as the configuration of the adaptive jitter buffer are copied to the new stream. In this way, unnecessary initialization and a miss-match in the size of both jitter buffers and in the RTP sequence numbers and timestamps can both be avoided.

During a changeover, both of the Media Streams are connected to the signal bridge, but only the original stream is attached to the Media Transport. A dedicated Media Transport, and so a dedicated RTP socket, is not created for the new Media Stream to avoid any potential firewall problems that might occur in the network and interrupt the transmission [17]. Instead, both of the streams communicate via the RTP Sessions with each other, and the RTP session of the original Media Stream is used to forward the data of the new Media Stream, which is based on the detection of the payload type of the RTP packets [18]. This design also provides a higher

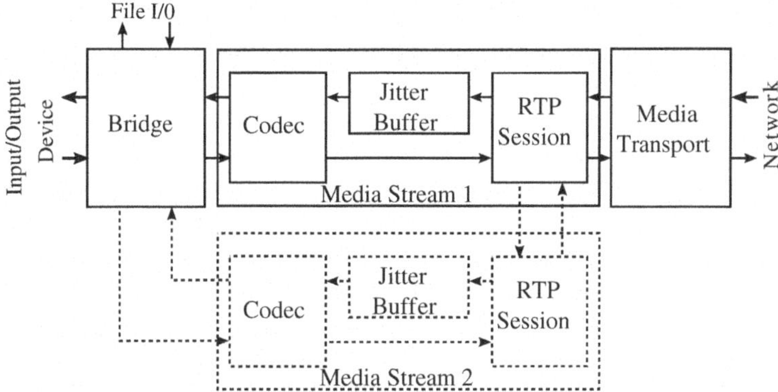

Fig. 3.5 Software architecture and media flow between the new (*dashed*) and the original (*solid*) Media Stream during a codec changeover

robustness of the solution. In case the new stream cannot be initialized properly, the initial stream can be used further.

The coexistence of two Media Streams during a changeover requires an additional synchronization of these entities to timely switch between the codecs and to finish the changeover procedure. The stream synchronization of the presented solution is realized in three steps:

- The forwarding of data in the new buffer is locked until a RTP timestamp match between both streams is detected.
- Immediately after the new stream successfully decodes the first signal frame, the old media stream is stopped, detached from the Media Transport and destroyed. Thereafter, the new Media Stream is attached to the Media Transport and acts as a single stream.
- As long as the decoding of the new payload type fails, the signal bridge distributes data that is received from the original stream only.

As a result, the presented implementation enables to decode packets that are stored in the jitter buffer of the original stream, as long as the newly created jitter buffer is not filled up with packets and the new codec does not provide a decoded signal. Moreover, data loss can be avoided because packets of both payload types can be processed. However, according to the above described synchronization steps both encoders might be active until the new decoder is not fully initialized, which may lead to increased transmission bit rate and power consumption during a changeover. This is in particular the case when video codecs are used. These codecs introduce certain processing delay, and the decoded signal cannot be provided until the codec is fed with a sufficient amount of data. As a result, the transmission bandwidth during a changeover might be increased, which leads to a violation of the last requirement from the list that is presented in the previous section. However, this design drawback was accepted in lieu of precise signal synchronization and of improved user experience. If immediate decoding of data is supported by the new codec, the technique will avoid any data duplication by default. Finally, the primary condition is fulfilled and the probability that at least one of both jitter buffers contains a sufficient number of packets is guaranteed, provided that the necessary condition $p_r = 1$ is satisfied for both codecs.

As a contrast scheme to the above approach, a baseline switching technique was implemented. The simplistic approach was designed without any consideration of the aforementioned requirements of a seamless codec changeover. Therefore, after the new codec is re-negotiated the old Media Stream is immediately destroyed and it is replaced by a new entity. As a result, the set S_i of data buffered in the initial stream is lost, packets of the old codec are rejected due to an unknown payload type, and the pre-buffering phase of the new stream cannot be bypassed. Finally, the condition $p_r = 1$ is violated, because n_j and p_n are increased, and an unavoidable gap in the decoded signal is perceived by the users. However, it is emphasised that this solution was used as a baseline for benchmarking the actual algorithm that was presented above.

3.3.3 Qualitative Evaluation

The quality of a signal that is generated during a codec changeover was evaluated in the domain of speech and video separately. The quality of the speech signal was instrumentally estimated with the PESQ model, and the results that were collected for both of the changeover techniques were compared with each other. It has to be emphasised that PESQ as such was not designed to cope with degradations due to a codec changeover and a valid assessment of the signal quality in this case is only possible by conducting a user test. However, because PESQ can cope with missaligned signals that can in particular occur during a changeover, and because only narrowband codecs were used in this evaluation, the instrumental methodology was selected to reduce the complexity of the benchmark. In the video case, an instrumental measurement of the video signal quality was skipped, because signal-based metrics, such as PSNR or SSIM, cannot cope with miss aligned reference and degraded signals. Therefore, the quality of the signal during a video codec changeover was analysed based on a subset of results from the user tests that were conducted in scope of this work. A direct comparison to the baseline implementation was skipped in this case, as the main performance differences have been already identified in the speech context.

To evaluate the signal quality during a speech codec changeover, a set of 80 test files was generated. The source samples were approx. 8 s long and consisted of two spoken sentences that were separated by a short pause, which is compliant with the requirements that are provided in the PESQ specification [19]. The test material was produced by streaming the speech data from the Correspondent to the Mobile Node in the testbed. Each transmission was initiated with the GSM Full Rate (12.65 kbit/s) codec that was replaced with G.711 (A-law PCM, 64 kbit/s) after approx. 1.5 s from the transmission begin. The codec was always switched together with the network,

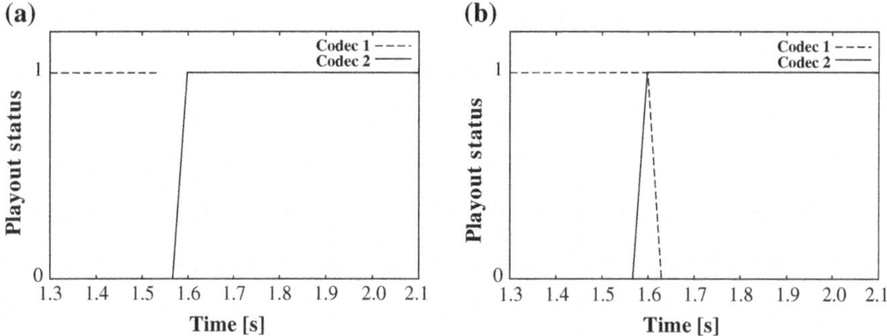

Fig. 3.6 Play-out status of the jitter buffer during a codec changeover. **a** Baseline algorithm. **b** Proposed algorithm

and a handover between WLAN and HSDPA was executed directly before or directly after the codec changeover. The combination of both procedures was applied due to the initial assumption of this work (cf. Sect. 4.4.1) that when the network is switched the speech codec has to be adapted to new transmission circumstances.

The quality estimates that were obtained using the PESQ model were averaged per condition and compared, showing an averaged quality improvement of 0.3 MOS when the proposed technique was used. This is a significant difference that was confirmed by a two-sided t-test ($t = -3.787$, $p < 0.001$).

Further analysis of the test material revealed that in 90 % of the test cases when the proposed switching technique was used, the playback interruption could be avoided. For illustration, a comparison of the play-out timeline that was measured at the output of the jitter buffer for both changeover techniques is depicted in Fig. 3.6. It is shown, that the baseline changeover technique always interrupted the signal playback, which is indicated by an abruptly ending status curve of the first buffer and by the initialization period that is observed before data from the new buffer is forwarded. This play-out gap is reduced when the primary technique of codec changeover is used. In this case, the first codec decodes the buffered data until the first buffer becomes empty, and simultaneously the initialization of the new buffer is bypassed.

The evaluation of the signal quality during a video codec changeover was based on a subset of conditions from Test V1 and V2 which included a changeover between H.264 and MPEG-4 at 256 or 1536 kbit/s in a loss free WiFi channel (cf. Fig. 3.7). The observed effect was consistent across both tests when the encoding bit rate was low, but it was different when the video bit rate was high. When the video bit rate was low, the main expectation of the analysis was fulfilled and the quality of a signal that was affected by a changeover was rated approx. between the reference quality that is provided by both codecs. In this case, the end quality rating was lower when it was switched to MPEG-4, which, however, is related to the fact that MPEG-4 at 256 kbit/s provides substantially lower video quality than H.264 at the same bit rate, and a degradation of quality is perceived stronger than a quality improvement (cf.

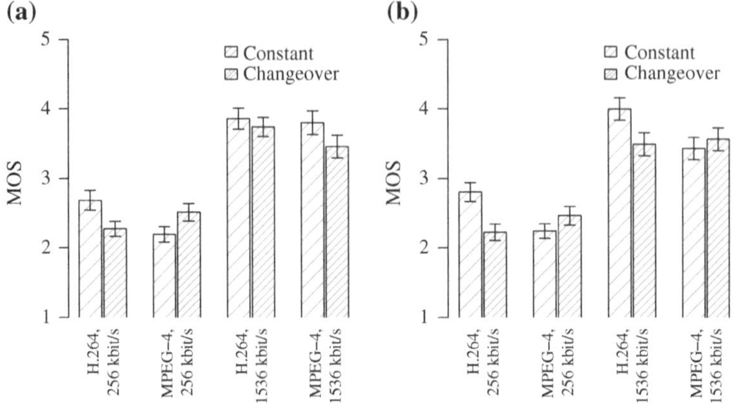

Fig. 3.7 Judgments of selected codec changeover conditions in **a** Test V1 and **b** Test V2. MOS_v and $MOS_{v|av}$ with 95 % CI, respectively. The x-axis describes the initial connection state, and the changeover was applied between two video codecs configured to the same bit rate

Sect. 5.2.2). When the video bit rate was high, the trend was inconsistent and it was violated in Test V1. In this case, video quality was always degraded below the reference quality that was provided by both of the codecs. When it was switched to MPEG-4 the impairment was negligible, with the end quality rating close to that of MPEG-4. However, when the codec was switched in the opposite direction, the video quality was significantly degraded, which was an unexpected result. Further analysis of this test condition revealed that in the utilized implementation of H.264 the initialization of the decoder required more time than that in case of MPEG-4. Therefore, when the codec was switched and until the H.264 decoder was initialized, data encoded by both codecs was transmitted in a duplicated manner. As a consequence, the wireless interface was overloaded, and packet loss occurred. As a result, the aforementioned requirements of a seamless codec changeover that were described in this chapter were violated, which indicates that a precise synchronization of video frames at high transmission bit rates might have a negative effect on the video signal, in particular when a long period of time is required to initialize a video decoder. Other solutions, such as a video freeze might be considered in this case to improve the user experience and to increase the scalability of a system. Finally, the difference in this particular case that was observed between Test V1 and V2 was related to different content types that were analysed in both of the tests. Less dynamics of the source material that was used in Test V2 resulted in less visible artifacts due to a changeover. This indicates that in the current shape the proposed technique of changing between video codecs is more suitable for video telephony than for streaming of general content types.

The technique presented in this section enables to switch the speech and video codecs during an ongoing session. The transition is seamless and the active session is not interrupted. Moreover, the decoding gap is reduced as well as possible through the implementation of changeover requirements that were defined in this section. If all of them are considered, the quality of a changeover can be significantly improved as compared to a case when a simplistic switching approach is used. However, it was revealed that the proposed design and precise synchronization of video decoders that require long codec initialization times might lead to visible artifacts when high bit rate video is provided. This showed that the proposed solution is more suitable for video telephony where less content dynamics are expected than for any other systems. In the latter case, other techniques such as freezing of the video play-out until the new decoder is active might be a more resource friendly solution to mask the visible impact of a codec changeover.

3.4 Testbed Controller

The number of components that are integrated in the testbed and hence the number of capabilities of the setup increased the complexity of the system control. To solve this issue and to enable automated data processing in the testbed, an additional control unit has been developed and integrated in the test client on the Mobile Node. The test controller consists of three core elements that provide basic functionality, and of several satellite modules that are implemented to control the individual modules of the testbed (cf. Fig. 3.8).

The core components of the test controller are: a Configuration Module, an XML Module, and a Thread Pool. The Configuration Module is used to adjust the testbed settings that are loaded from a configuration file. Those are for example the IP addresses of the testbed nodes, or a list of available network interfaces which is represented as a hash of interface names in the system (e.g. wlan0 or ppp0) and human-readable network names (e.g. WiFi or HSDPA). Next, the XML module is used to read and process XML files that define the test protocols. Finally, the Thread Pool is implemented to initialize and manage a group of application threads that are used on-demand in the setup to execute the testbed routines. All the processes are launched through the control modules that are attached to the core unit. The modules are subdivided according the their functionality as follows:

- *lmip Controller* is used to schedule a network handover.
- *netem Controller* is used to artificially degrade the link quality of the HSDPA link.
- *Router Lab Controller* is used to control the link impairment generated on the Dummynet node in the Router Lab.
- *Codec Controller* is used to schedule a speech or a video codec changeover.
- *tcpdump Controller* is used to record network traces and save them to disk.
- *Remote Sender Controller* is used to start and control the test client on the Correspondent Node.
- *Recording Controller* is used to record the audio and video signal to files.

As already mentioned above, the testbed supports automated processing of test protocols. To define a test, an event-based concept of test definition was developed in this work. Each change of a transmission parameter is considered as a single event,

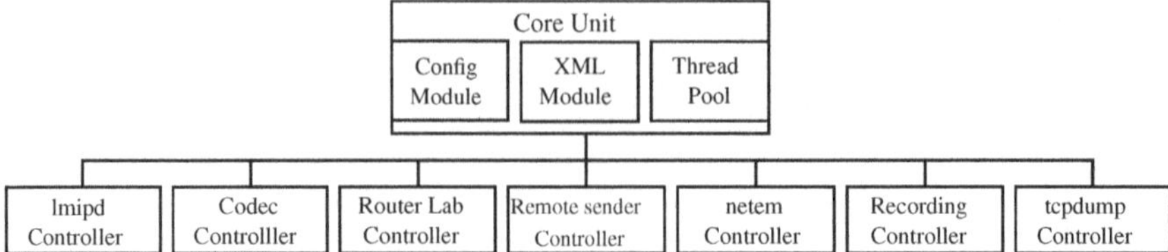

Fig. 3.8 Software components of the test controller

and each event includes a value and time that describes when the parameter should be changed. An example of an event that can be defined in the testbed is a network or codec changeover, as well as a change of the artificially generated packet loss. One or more events can be used to define a test condition, and one or more test conditions can be included in a single test. As a result, the tests and test conditions can be arbitrary defined, which includes test conditions that are affected by multiple changes of parameters, as well as conditions that represent constant quality.

```
<test_condition>
  <video_input>input.yuv</video_input>
  <video_output>output.yuv</video_output>
  <network>
    <value>wifi</value>
    <time>0</time>
  </network>
  <video_codec>
    <value>H264_1536</value>
    <time>0</time>
  </video_codec>
  <packet_loss>
    <value>3.5%</value>
    <time>2.5</time>
  </packet_loss>
  <video_codec>
    <value>MPEG4_1536</value>
    <time>3</time>
  </video_codec>
</test_condition>
```

Listing 3.1 Example of a test case definition

An example of a test condition is presented in Listing 3.1. The XML description consists of an input and output filename, and of four test events: one network, one packet loss, and two video codec events. Each of the events is specified by its name, and it includes a value of the new setting and time when the change should be scheduled. As a result, the processing cycle of this test condition is as follows. Before the transmission, the initial parameters are configured (time 0 s). The network is configured to WiFi and the video codec is switched to H.264 at the bit rate of 1536 kbit/s. Then, (which is not specified in the test case) the transmission along with background processes, such as network tracing and video recording, are started. 2.5 s later the connection quality is artificially degraded by 3.5 % of packet loss, and half a second later (time 3 s), the video codec is switched to MPEG-4. The transmission is automatically terminated after the source file is completely transmitted, and all of the related processes that are running in the background are stopped.

As a result, the testbed controller that was presented in this section enables to centralize the control of the setup and to automate the process of test data generation. Moreover, the event-based concept of test definition allows to generate any amount of data that is required to investigate the quality of arbitrary transmission scenarios. This includes both, stable and dynamic conditions, in which the streaming properties are changed several times during an ongoing session.

3.5 Demonstrator

Frequently, the signal quality that occurs under certain circumstances has to be visualized. For this purpose, the capabilities of the testbed were reused, and a demonstration system has been developed. The main component of the presented demonstrator is a graphical user interface that is loaded in the test client and that makes use of the test controller API.

The demo GUI is implemented based on QT, SDL library, and a set of widgets that are distributed with the Embedded Widgets Demo [20]. The interface (cf. Fig. 3.9) consists of several control and status widgets that are placed around the centrally located window of video display. The control widgets are located at the bottom of the screen. They are sorted in groups according to the parameters they control. In the first group, the codec control is provided. A set of buttons to switch between the audio and video codecs and switches to toggle the Packet Loss Concealment (PLC) or Voice Activity Detection (VAD) are available. The name of the active codec is highlighted in green. A similar structure is used in the second group of control widgets to control the networking conditions. The name of the active network interface is highlighted in green, and a set of widgets and sliders is provided to switch between the networks and to control the artificial degradation of the connection quality. Packet loss, delay or delay variation can be adjusted. In the last two control boxes that are located in the bottom of the screen, a gauge that shows the status of audiovisual synchronization and a switch to start the demonstration are provided.

The status information about the active media streams is presented on the sides of the video display. The audio widgets are placed on the left, and the video widgets on the right side of the screen. Both groups include a gauge and a slider that show the streaming bit rate and the size of the jitter buffer, including the low and high boundary in the latter case. Finally, the lamps that are located on top of the screen provide additional status information about the signal decoding. Any interruptions, errors, or use of PLC are notified by flashing red.

As a result, the above described demonstrator enables to experiment with various transmission conditions and simultaneously to visualize the achieved effects. All of the described widgets can be used at will during the demonstration, and the resulting signal quality and status of the application can be observed in real-time. It has to be

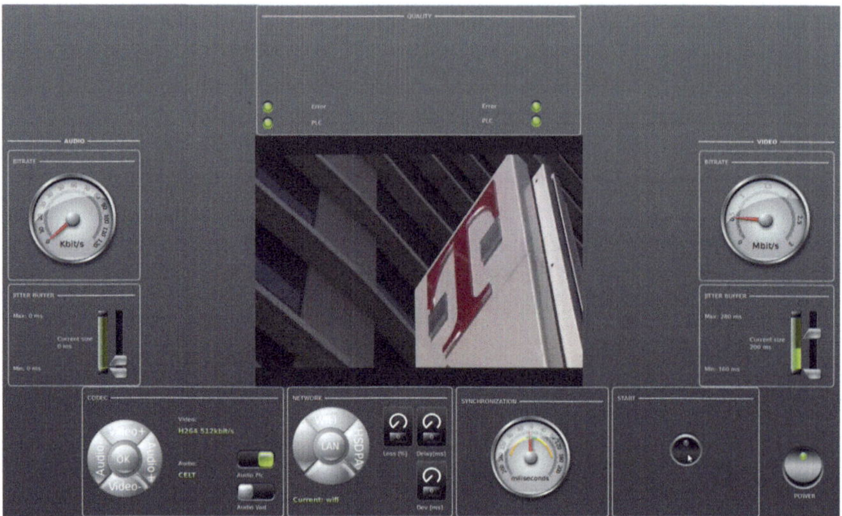

Fig. 3.9 User interface of the demonstrator

emphasised that the demonstrator was implemented on top of an existing testbed, which shows that testbeds can be reused for multiple purposes.

3.6 Chapter Summary

The testbed that was presented in this chapter enables to evaluate user experience of speech and video telephony in heterogeneous wireless networks. It consist of a network infrastructure, of a speech and video telephony client, and of a testbed controller. The networking setup is assembled of multiple IP nodes that are interconnected through several heterogeneous wireless technologies. Roaming between these networks is supported by Mobile IP, and networking degradations due to packet loss or variable network delay or background traffic can be generated. All this features enable to use the testbed for research on user perception of quality in heterogeneous networks.

More capabilities of the assembled setup are provided by the test application that was deployed to enable speech and video telephony in the testbed. The application is based on a stand-alone VoIP project that has been extended in several ways during this work. The video support has been implemented, and an additional set of features that are necessary to research time-varying quality and optimization on cross-layer parameters has been integrated. The core feature that was designed during this work is a technique to switch the speech and video codecs during an ongoing transmission. The transition between the codecs is seamless, an active session is not interrupted, and signal decoding artifacts are reduced as far as possible. To achieve this goal a set of requirements that a seamless codec changeover has to fulfill was designed. Moreover, a multitude of codecs has been integrated in the application, and thus diverse speech and video coding configurations can be evaluated. The perceptual trade-offs between the audio and video quality can be explored, and in particular, strategies for adaptation of the perceived quality during an ongoing transmission can be developed. More features of the testbed that are related to the optimization of real-time streaming in heterogeneous networks result from the reconfigurability and monitoring of the de-jittering buffers that are supported in the application. Experiments with different sizes of the buffers can be performed, and the buffering status, in particular data loss that might occur in the application, can be examined.

The flexible design of the test client also enables to use it for various types of tests, such as interactive and non-interactive tests. For the interactive tests, peripheral devices, such as cameras and headsets, can be used, as well as various user interfaces that are implemented as shared libraries and that are compatible with the client. As a matter of fact, this way the application can also be used to test diverse user interfaces. For non-interactive tests that are based on test material that is generated offline, test data can be generated. This process is automated through the implementation of the testbed controller, and the test protocols can be flexibly defined. The event-based concept of test definition allows to generate any amount of data that is required to investigate the quality of arbitrary transmission scenarios. This includes both, stable and dynamic conditions, in which the streaming properties are changed several times during an ongoing session. During the recording, any effects that would be perceived by the users, such as video freezing and decoding artifacts, and rescaling, can be captured, as well as network and jitter buffer traces can be recorded.

The development of the presented testbed was a complex process that included several evolution stages. The modifications of the testbed were motivated by the increasing requirements of the research project as the work evolved. Therefore, during the initial stage of the project, the original VoIP functionality that is provided by PJPROJECT was used and the first prototypes of the codec changeover technique were developed [6, 21]. In the next stage of the project that addressed the quality of video calls, video support was added, the codec changeover technique was refined, and more codecs have been integrated [22]. Finally, the complexity of the testbed control was reduced by the development of the testbed controller and by the definition of the event-based concept of test definition [5]. As a consequence, in the final shape the testbed was used in several research projects that addressed user perception of audiovisual quality in telephony systems. Moreover, the presented testbed was also reused to create a demonstration environment in which the transmission phenomena that were addressed in the project could be presented.

References

1. Klaue J, Rathke B, Wolisz A (2003) Evalvid-A framework for video transmission and quality evaluation. Comput Perform Eval Modell Tech Tool 255–272
2. Kellerer W, Steinbach E, Eisert P, Girod B (2002) A real-time internet streaming media testbed. In: Proceedings of IEEE international conference on multimedia and expo (ICME). Lusanne, pp 453–456

3. Li F, Chung J, Li M, Wu H, Claypool M, Kinicki R (2005) Network and link layer measurements of streaming video over a wireless campus network application. Passive Active NetwMeasur pp 189–202
4. Dutta A, Chen JC, Das S, Elaoud M, Famolari D, Madhani S, McAuley S, Tauil M, Baba S, Maeda T, Nakajima N, Ohba Y, Schulzrinne H (2001) Implementing a testbed for mobile multimedia. In: Proceedings of the IEEE global telecommunications conference (GLOBECOM). San Antonio, pp 1944–1949
5. Lewcio B, Möller S (2011) A testbed for QoE-based multimedia streaming optimization in heterogeneous wireless networks. In: Proceedings of IEEE international conference on signal processing and communication systems (ICSPCS). Honolulu, pp 1–9
6. Vidales P, Kirschnick N, Steuer F, Lewcio B, Wältermann M, Möller S (2008) Mobisense testbed: merging user perception and network performance. In: Proceedings of international ICST conference on testbeds and research infrastructures for the development of networks and communities (TRIDENTCOM). Innsbruck
7. Mehmood A, Hohlfeld O, Levin D, Wundsam A, Ciucu F, Schneider F, Feldmann A, Braun RP (2010) The routerlab - emulating internet characteristics in a room. In: Proceedings of ITG symposium on photonic networks. Germany, pp 1–8
8. Jung J-W, Mudumbai R, Montgomery D, Hyun-Kook K (2003) Performance evaluation of two layered mobility management using mobile IP and session initiation protocol. In: Proceedings of the IEEE global telecommunications conference (GLOBECOM). San Francisco, pp 1190–1194
9. Net:Netem - The Linux Foundation (2009) http://www.linux-foundation.org/en/Net:Netem. Accessed January 20 2009
10. Website of pjsip, an open source SIP stack and media stack for presence, instant messaging, and multimedia communication (2009) http://pjsip.org/. Accessed January 20 2009
11. Rosenberg J, Schulzrinne H, Camarillo G, Johnston A, Peterson J, Sparks R, Handley M, Schooler E (2002) SIP: session initiation protocol. RFC 3261 (Proposed standard), Internet Engineering Task Force
12. Schulzrinne H, Casner S, Frederick R, Jacobson V (2003) RTP: a transport protocol for real-time applications. RFC 3550 (Standard), Internet Engineering Task Force
13. Website of ffmpeg. http://www.ffmpeg.org/. Accessed June 07 2008
14. Schmidt M, de Bont F, Doehla S, Kim J (2011) RTP payload format for MPEG-4 audio/visual streams. RFC 6416 (Proposed standard), Internet Engineering Task Force
15. Wang Y-K, Even R, Kristensen T, Jesup R (2011) RTP payload format for H.264 Video. RFC 6184 (Proposed standard), Internet Engineering Task Force
16. Baratvand M, Tabandeh M, Behboodi A, Ahmadi AF (2008) Jitter-buffer management for VoIP over wireless LAN in a limited resource device. In: Proceedings of international conference on networking and services. Washington, pp 90–95
17. Rosenberg J, Schulzrinne H (2002) An offer/answer model with session description protocol (SDP). RFC 3264 (Proposed standard), Internet Engineering Task Force
18. Lewcio B (2007) Assessment and adaptation of voice quality in All-IP heterogeneous networks. Diploma thesis (unpublished), Hochschule für Telekommunikation. Leipzig
19. ITU-T Rec. P.862 (2001) Perceptual evaluation of speech quality (PESQ): An objective method for end-to-end speech quality assessment of narrow-band telephone networks and speech codecs. In: International telecommunication union, Geneve
20. Website of embedded widgets and demos (2011) http://developer.qt.nokia.com/wiki/EmbeddedWidgetDemos. Accessed June 14 2011
21. Wältermann M, Lewcio B, Vidales P, Möller S (2008) A technique for seamless VoIP-codec switching in next generation networks. In: Proceedings of IEEE international conference on communications (ICC). Beijing, pp 1772–1776
22. Lewcio B, Möller S, Vidales P (2011) A method for seamless codec changeover during active video calls. In: Proceedings of IEEE computer-aided modeling analysis and design of communication links and network (CAMAD). Kyoto, pp 133–137

Chapter 4
Experience of Speech Telephony

The quality of ongoing Voice over IP (VoIP) calls in heterogeneous all-IP networks is expected to vary due to mobility and instability of link-layer conditions. Transmission impairments, such as packet loss, will occur, and therefore roaming between heterogeneous networks will be necessary to maintain a phone call. To this end, quality adaptation procedures will be required to adjust the transmission for the changing transmission circumstances. Speech codecs or network connections will be configured during run time to ensure maximal user experience of the telephony quality.

The analysis in this section is focused on the perception of quality in WiFi and HSDPA networks, in which various narrowband and wideband speech codecs can be used, and most importantly, where it can be switched between those networks and codecs during an ongoing connection. The user perception is explored based on a wide range of user tests. The test data is used to understand user experience and to draw guidelines for a perception-based mobility management. Furthermore, the data is also used to analyse how well the existing prediction models can estimate the perceived speech quality in the context of time-varying transmission in heterogeneous wireless networks, and some new approaches of instrumental quality assessment are introduced.

4.1 Experimental Design

In order to get analytic insights into user experience of VoIP quality, user perception of entire phone calls must be considered. Standard listening-only tests, such as ITU-T Rec. P.800 [1], which make use of speech samples with a length of approx. 4–8 s are not suitable for this purpose. On the other hand, conversational tests despite being comparable to normal telephone usage and thus being ecologically valid place a content-related focus on the users attention. In such a situation, users are generally less analytic in their judgments, and it might happen that subtle perceptual differences get blurred. As a compromise, to capture more decent differences in call quality a methodology that is based on the emulation of a call situation was recommended by ETSI in TR 102 506 v.1.2.1 [2]. This test protocol is not-based on a real conversation,

B. Lewcio, *Management of Speech and Video Telephony Quality in Heterogeneous Wireless Networks*, T-Labs Series in Telecommunication Services,
DOI: 10.1007/978-3-319-02102-7_4,

but creates a situation in which a test participant has to focus on the presented material. Content-related questions are asked during the test to enforce user interaction with the test system.

Despite the above diversity of test methodologies, in the context of time-varying transmission additional aspects of user perception have to be noted. It is necessary to understand how a single effect, i.e. a quality change, is perceived, but also how those single changes affect the overall experience of the call quality. As an example, how is a change of the speech codec perceived, and what is the effect of this single change on the end quality of a call? Therefore, to cover all of the above discussed aspects of quality perception, three types of user tests were conducted to address the entire VoIP quality perception in heterogeneous wireless networks. As a result, the tests were subdivided as follows:

- Three tests were focused on quality perception of emulated calls. Tests S1a, S2a, S3a were used to put a magnifier on the listening quality of VoIP service in a controlled situation that is close to a real phone call.
- Three tests addressed the quality of short samples. Tests S1b, S2b, S3b were used to understand how users perceive single effects that may occur in heterogeneous networks.
- One test was designed to validate the previous results and investigate the quality perception during real phone calls. Test S4 was an attempt to create a real phone conversation over heterogeneous networks under controlled laboratory conditions.

The first and second type of tests are frequently referred to as listening-only tests, and the last test is designated as a conversational test in this work.

4.1.1 Subjective Tests of Listening-Only Quality

Test procedures

The experiments that were based on the emulation of a phone call were designed according to the standard recommendation provided in ETSI TR 102 506 v.1.2.1. One experiment consisted of multiple emulated calls that were presented to the test participants. Each of the calls was constructed of five samples that were played-out in a meaningful order to tell a story that is typical for a phone call. After each of the first four samples was played-out, the test participants were asked to orally answer a content-related question. After the last sample of a call was presented, the overall call quality was judged on the 5-point MOS scale. The test samples and pauses for answering the questions were approx. 6 s long, which resulted in a total duration of 60 s per call emulation.

The quality of samples that were concatenated within a call created a call quality profile. Depending on whether the quality of the individual samples was identical or not, constant or irregular call quality profiles were possible. The profiles were designed in a way to address typical degradation patterns expected in heterogeneous networks. This includes for example users moving towards or away from an access point during a call.

The quality of the samples that were used in each of the emulated calls as well as of some additional material was judged in a separate test that was following the listening-only procedure described in ITU-T Rec. P.800 [1]. The approx. 6 s long test files were sequentially presented to the test participants who judged the quality of each sample on the 5-point MOS scale. The short samples were used to quantify the influence of heterogeneous technology on user perception, in particular addressing how the phenomenon of a technology switch during an ongoing transmission is perceived.

In the following, the tests that were based on the call emulation are identified as "a" tests, and all short sample tests are denoted as "b" tests. All "a" and "b" tests that belong together were always conducted during one test session. The test participants were invited to a sound-insulated laboratory where they were instructed about the purpose of the test, and where they listened to the test material in three sessions of approx. 25 min. each. Two sessions for the "a" parts, and one session for the "b" part. The test set-up was compliant with the requirements given in the respective standard recommendations. The test room characteristics followed that recommended in ITU-T Rec. P.800 and the test material was presented through a Sennheiser HMD 410 headset at a comfortable listening level with a background noise below 35 dB(A). Prior to each experiment the test participants were interviewed about any hearing impairments. Then, they followed a training phase in order to familiarize with the test procedure, degradation types, and the quality range included in each of the tests. The resulting tests 1 a/b, 2 a/b, 3 a/b were carried out at distinct points in time, which also implies different participant groups per test set. They were recruited from the normal telephone-user population for Test 1 and 2, and from a non-expert group of the university campus for all the other tests. The effort of completing a test was rewarded with a shopping voucher.

Source material

The source material that was used in all of the tests included two male and two female participants of a phone call who were recorded while accomplishing a certain task. For example, it was the goal of a call to order a washing machine, to rent a car, or to make an appointment with a doctor. Each of the story-tellings was split into 10 samples of approx. 6 s, which generated a pool of 40 samples that are typical for telephony. However, subsets of only 5 interrelated samples were used for each call emulation.

Test S1 and S2 shared one pool of source samples that were stored in an uncompressed wideband format. The selection of wideband-filtered material was determined by the first commercial roll-outs of wideband telephony systems at the time the tests were carried out. Test S3 included similar content, but it was recorded separately in a studio quality without any bandwidth limitations to allow research on the quality of super wideband VoIP.

Test conditions

The first two user tests that were carried out in this work, Tests 1a/b and 2a/b, were designed to identify the main perceptual effects that may occur in heterogeneous wireless networks. Network handover, changeover between narrowband and wideband codecs, and packet loss were the addressed mobility phenomena. The initial tests followed an exploratory strategy instead of a systematical analysis due to several reasons. First of all, the addressed phenomena of time-varying quality in heterogeneous networks were not fully explored before, and the main effects had to be identified. However, an analysis of all of the potential test cases within a single test was not possible, because the number of test conditions was too large. A particular difficulty arised from the number of potential switching combinations, and thus only anchor conditions that were relevant to focus on time-varying quality were selected.

As a consequence, Test 1a was a general test that addressed the user perception of transmission changes in heterogeneous networks, and for this purpose the following call quality profiles were designed:

- 2 reference profiles that represented unimpaired wideband and narrowband speech transmission in WiFi and HSDPA, respectively.
- 4 profiles that targeted the phenomenon of switching between wideband and narrowband speech quality at the beginning (2nd sample), or at the end (4th sample) of an emulated call.
- 4 profiles that emulated a user moving away from an access point by increasing packet loss until the middle of a call (3rd sample) and eventually a handover to a loss-free network that requires a codec switch.

All of the profiles are depicted in Fig. A.1 and described in Table A.2. The test material was encoded with G.722.2 at 23.05 kbit/s and G.711 at 64 kbit/s. Both of the codecs were selected according to the ITU-T Rec. G.113 [3] as those providing the reference narrowband and wideband speech quality.

To put a magnifier on the interesting results of the first test, Test S2 was designed. This test was focused on the user perception of handovers to a loss-free and narrowband quality network due to packet loss in wideband. Moreover, this test also addressed multiple changes of the speech bandwidth during a single call, which resulted in the design of the following call quality profiles:

- 2 reference profiles that represented unimpaired wideband or narrowband quality in WiFi and HSDPA, respectively.
- 2 profiles that addressed the phenomenon of switching from narrowband to wideband as a potential call quality improvement at the beginning and, in opposite to Test 1a, in the middle (3rd sample) of a call.
- 3 profiles that were focused on the emulation of users moving away from an access point that were affected by lower packet loss rates than in Test 1a and by a handover to loss-free HSDPA with narrowband codec.
- 2 profiles that were created to compare the user perception of a period of narrowband quality in the middle of a call to the same period of wideband transmission affected by packet loss.
- 1 profile that introduced multiple network and codec switches.

All of the profiles are depicted in Fig. A.2 and described in Table A.6. The speech coding of these profiles was realized with G.722.2 at 12.65 kbit/s and G.711. A lower bit rate of G.722.2, as compared to the previous test, was used to compare if the perception of the speech bandwidth switching will change, when the coding quality difference is smaller.

The design of Tests S1 and S2 turned out to have two limitations. It was driven by the internal assumption that if a network is switched it will be required to adapt the speech codec to new networking circumstances as well. Second, the tests were focused on narrowband and wideband quality only. To explore the perception of a codec changeover in a multi band system in which narrowband, wideband, and especially super wideband speech codecs coexist, Test S3 was designed. The investigation of network handover and packet loss was omitted in this test to quantify the impact of an isolated codec changeover on user perception only. Therefore, loss-free WiFi was used in this test.

As a result, the design of Test S3 followed a more systematic strategy, as the main effects were already identified during previous tests and a more detailed outcome was expected. Therefore, the data set of Test S3a consisted of the following profiles:

- 3 reference profiles that represented unimpaired narrowband, wideband, and super wideband quality.
- 6 profiles that addressed user perception of changeover between wideband and super wideband codecs, at the same time addressing the impact of the temporal position (beginning, middle, end of a call) of a switch.
- 2 profiles that included a changeover between narrowband and wideband codecs in the middle of a call, which was an anchor to the previous tests.
- 2 profiles that introduced a changeover between narrowband and super wideband codecs, as a comparison to the above two test profiles.
- 2 profiles that addressed switching between narrowband and super wideband codecs with switching to wideband quality as an intermediate step to smoothen the quality jump.

All of the profiles are depicted in Fig. A.3 and described in Table A.9. The test material was encoded with G.711, G.722, and G.722.1C that were applied to the narrowband, wideband, and super wideband signal, respectively. They were selected for the test because all of them operate at similar bit rates, but they provide different bandwidth of the speech signal.

The design of the (short sample) "b" tests in Test 1, 2, and 3 contained all conditions that were used as components of the corresponding emulated calls, as well as some additional samples with degradations related to each of the tests. The test conditions were grouped into stable and dynamic conditions. While stable conditions represented constant streaming characteristics, the dynamic conditions included a change of transmission properties in the middle of a test sample. Accordingly, network handover and speech codec changeover, or combinations of both were applied as switching events relevant for heterogeneous networks. Both of the condition groups were generated either in a clean transmission channel or in a channel affected by packet loss, to explore quality adaptation strategies under diverse network circumstances. The detailed lists of test conditions in Test S1b, S2b, S3b are presented respectively in Tables A.4, A.8, A.11 in the appendix of this book. It has to be emphasised that it was one of the main initial assumptions of the presented test design that if a network is switched it will be required to adapt the speech codec to new networking circumstances. Therefore, in most of the test cases the codec changeover and network handover were performed together as assisting procedures. This assumption helped to limit the number of test combinations during the initial tests, which was necessary to discover the main effects. However, at the later stage of this work, in particular presented in the next chapter, this restriction was released, because the main effects were known and the concept of IP communication allows enough flexibility to decouple the network and codec selection.

4.1.2 Subjective Test of Conversational Quality

Test procedure

The test procedures described above addressed the quality perception in a listening-only situation. To analyse the quality during a real conversation, a separate test was conducted. The conversational Test S4 was compliant with the methodology proposed in [4]. It was based on real phone calls that were performed between two test participants in a laboratory environment. During each of the calls, the test subjects were jointly solving a task that was similar to an every-day conversation. For example, they had to order a ticket for a theater show, or make an appointment with a doctor. Each of the test participants acted one of the roles, either a caller or a callee and each of them possessed a piece of information that was required to solve the task. For example, the caller wanted to make an appointment with a doctor at a date that was suitable for him, but only the callee could see the list of available slots. Thus, essential information to solve the task was exchanged during the conversation.

The duration of a test call was limited to maximal 180 s. This restriction was of particular relevance because the time to solve a task was dependent on the performance of the test participants. Some of them aimed at quick solving of tasks, while some others preferred long conversations. To achieve a comparable length of the conducted test calls, the duration was artificially limited to a time slot that made leisured task completion possible. The test subjects were informed about the limitation of the call duration and a small count down clock was visible during the test (cf. Fig. 4.1). In case a call was automatically disconnected, the test participants were instructed to focus on the quality judgement only, even if the task could not be accomplished. The quality judgements were collected on the 5-point MOS scale.

The test setup consisted of two sound-isolated rooms that were conform with the ITU-T Rec. P.800 standard, and of a control room for the test supervisor. Each of the test participants used a laptop that represented either the Mobile or to the Correspondent Node of the Mobisense testbed (cf. Chap. 3 for details). Each of the laptops was equipped with a Sennheiser HMD 410 headset and an Edirol UA-25EX sound card as an audio input and output device. Moreover, the test VoIP application of the Mobisense testbed was additionally extended with a graphical user interface of a soft phone that was installed on the test computers (cf. Fig. 4.1). To initiate a phone call, the callers had to dial a certain telephone number that was assigned to

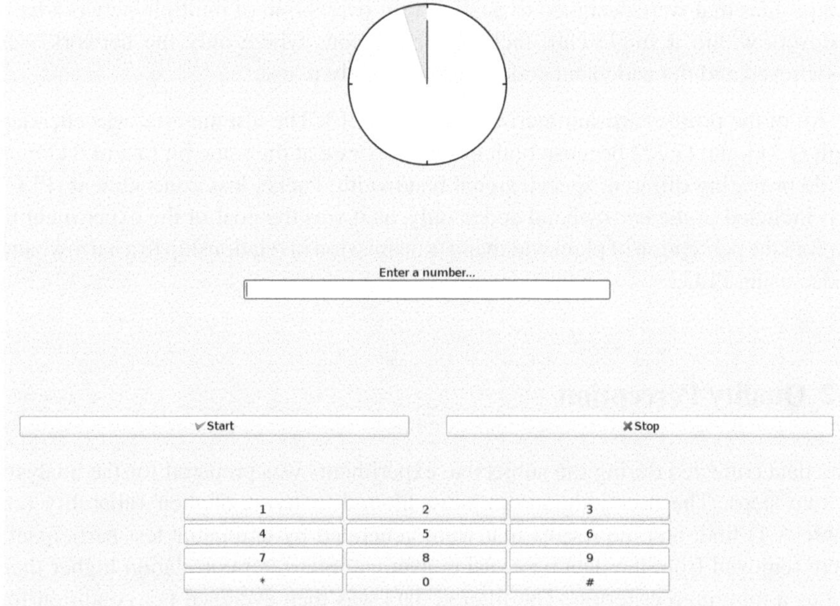

Fig. 4.1 User interface of the VoIP client during Test S4

a scenario. Therefore, to order the theater tickets users dialed a different number as when they made an appointment with a doctor. The numbers were checked for their correctness. When miss-typed, an invalid number message appeared on the screen. The test scenarios were randomized with the call quality profiles using the Graeco-Latin-Square-Designs method as described in [4].

Test conditions

Test S4 was designed to address the same main effects as the listening-only tests. However, due to a longer duration of the test calls, longer quality profiles were used prevailing profiles that included multiple transmission changes within a single call. Each test condition was used only once for each pair of test participants, and thus each of them performed 14 phone calls that were constructed of the following quality profiles:

- 2 reference profiles that represented unimpaired wideband, and narrowband quality in WiFi and HSDPA, respectively.
- 3 profiles that addressed how users perceive a handover to a loss-free narrowband network due to packet loss in a wideband network.
- 2 profiles that were focused on the perception of packet loss in wideband at the end of a call. Those conditions were used for comparison with the former 3 profiles.
- 3 profiles that were included to investigate how different duration and different degradation types in the middle of a wideband call is perceived. Packet loss or narrowband coding were in scope of work.

- 5 profiles that were designed to analyse user perception of multiple network handovers within a single call, including conditions where only the network was switched and the wideband codec was constantly in use.

All of the profiles are summarized in Table A.13. The test material was encoded with G.711 and G.722 because both of them operate at the same bit rate of 64 kbit/s while providing different speech signal bandwidth. Packet loss concealment (PLC) was included in the narrowband codec only, as it was the goal of the experiment to explore the perception of plain wideband transmission in relationship to a narrowband codec using PLC.

4.2 Quality Perception

The data collected during the subjective experiments was prepared for the analysis in two steps. The user judgements were filtered in terms of their reliability (cf. Table A.1) first, and the results that were generated by unreliable test participants were removed from the data base, using average inter-rater correlation higher than 0.3 as a minimum criterion. The filtered data was then averaged per condition for the further evaluation; the detailed results of all speech quality experiments can be found in Appendix A.1.

The analysis in this section is organized similarly to the structure that has been already introduced in the preceding section. The quality of the emulated calls is evaluated first, and next the analysis of the short samples quality is described, and the study of the quality perception during real calls is presented as last. Cross-test observations are provided chronologically at the relevant positions in the text as the evaluation evolves, as well as in the summary of the entire chapter.

4.2.1 Quality of Emulated Calls

The quality judgements that were collected during Test S1a (cf. Fig. 4.2) revealed that pure wideband transmission was rated best (#2) and substantially better than pure narrowband quality (#1). However, this relationship was changed when degradations due to packet loss occurred. If packet loss was increasing within of a call, the speech quality provided by the narrowband codec was rated considerably higher than that when the wideband codec was used (#7, 8), which was related to a higher packet loss robustness of the narrowband codec [5]. Moreover, in any case when the transmission was affected by packet loss, the quality was judged worst in the entire test. This observation reveals that packet loss was the most dominant factor of call quality degradation.

If the increasing packet loss in a call was eliminated by performing a handover to a loss-free network in the middle of an emulated call, which simultaneously required a

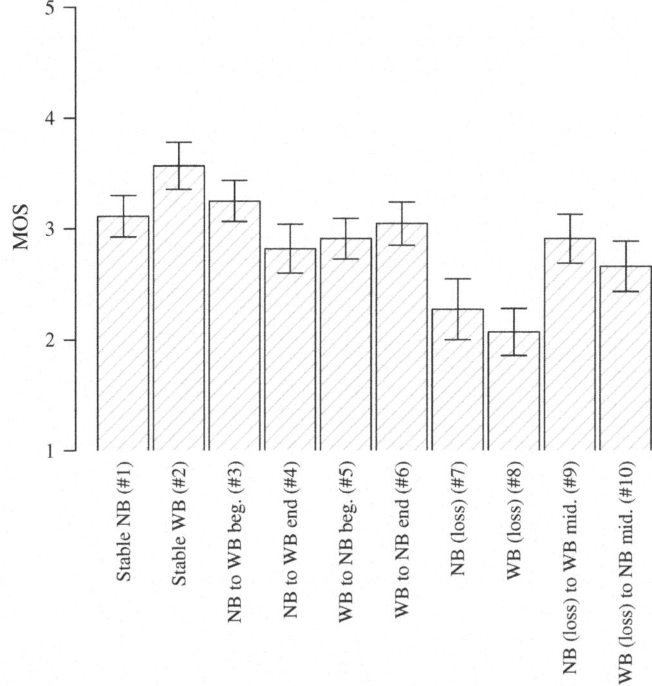

Fig. 4.2 User judgements of call quality in Test S1a. MOS_a, and 95 % CI

switch of the speech signal bandwidth, the perceived quality was always substantially improved (#9, 10). This finding held when the network and codec were changed in both directions (either from wideband to narrowband or from narrowband to wideband). However, the improvement was lower if the narrowband codec was used at the end of a call. Therefore, if packet loss occurs, early switching to a loss free network is advantageous in any case, even if it is required to reduce the bandwidth of the speech signal. However, the end quality experience was always lower than that of pure narrowband transmission, which is the first indication that, if a network and a codec switch can be foreseen, and at the same time a stable narrowband link is available, delivery of low, but stable narrowband quality improves the Quality of Experience.

When no packets were lost, raising the speech bandwidth from narrowband to wideband was only advantageous if a sufficient duration of a call remained in order to take profit of the improved quality (#3). If the switch to wideband was performed late in a call, the overall quality experience was degraded below that of pure narrowband (#4), which is related to degradation of short-term perception through a codec switch, which is addressed in the next section. In turn, decreasing the speech bandwidth from wideband to narrowband always degraded the call quality below the rating of pure narrowband, and the end quality was the lower, the earlier in a call it was switched

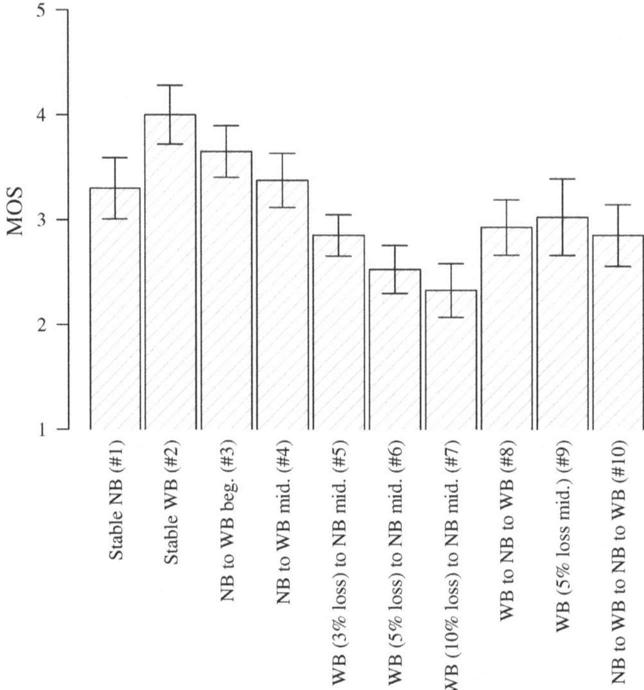

Fig. 4.3 User judgements of call quality in Test S2a. MOS_a, and 95 % CI

(#5, 6). As a general consequence, once narrowband occurred in a phone call, the end quality judgement was close to that of pure narrowband transmission. Thereby, the direction and temporal position of a switch had only a marginal impact on the user perception, which means that if packet loss does not affect the quality, switching between wideband and narrowband speech in most of the cases is not recommended. This is yet another argument that if a switch can be foreseen and a stable network that provides narrowband speech quality is available, this network should be used. Sticking to pure narrowband also improves the scalability of a transmission system, as the number of the mobility management operations is reduced.

The main findings from Test S1a were confirmed by the results of Test S2a (cf. Fig. 4.3), but also an anomaly in the wideband quality rating was disclosed. Even if a lower quality mode of the wideband codec was used in the second test, pure wideband was rated higher than in the first test. This effect might be related to the fact that decent differences within one speech bandwidth get blurred when wideband and narrowband speech are analysed together, but more tests are necessary to confirm this hypothesis. Nevertheless, Test S2a confirmed that pure wideband quality was rated best (#2), and that packet loss had the strongest impact on the speech quality (#5, 6, 7, 9). Moreover, it was confirmed that switching from narrowband to wideband is advantageous only if it occurs early in a call and if the wideband part is long (#3, 4). So, switching to wideband at the beginning of a call resulted in a quality that is rated approx. in the middle between the judgements of pure narrowband and of pure wideband. However, switching to wideband around the middle of a call resulted in a comparable score as keeping narrowband for the entire call. As a result, direct raising of the speech bandwidth from narrowband to wideband is a sensible instrument for quality adaptation that is efficient at the beginning of a call only. This finding leads to a new research question, whether the perception of a switch between wideband and narrowband can be improved by smoothening of this process and extending the duration of the transition phase [6]. To this end, the technical realization of this masking procedure is another key question. Potentially, adaptive bandpass filtering, artificial bandwidth extension, or longer fading of both signals through simultaneous transmission of both streams during the codec changeover may be considered.

Moreover, Test S2a also confirmed that a combined network and codec switch as an instrument for combating packet loss is perceived worse than pure narrowband transmission, and no profit was taken of the degraded wideband period at the beginning of a call (#1, 5). Thus, if it can be predicted that packet loss in wideband will occur, a network and a codec switch will be required, and at the same time a stable narrowband connection is available, it is advantageous for the user as well as for the system to perform the entire transmission in a loss-free narrowband network.

Changing the speech bandwidth was also not beneficial in the middle of a call, until a certain threshold of quality degradation was reached. A comparison of two quality profiles in which the middle part of a wideband call was either affected by 5 % (#9) of packet loss or it was switched to a loss-free narrowband network for the same period of time (#8) revealed that the quality of those profiles was rated below pure narrowband, but the degradations due to packet loss were slightly more tolerated by the users. Moreover, frequent network and codec changeovers within of a loss-free call turned out to be rated worse than a single change between wideband and narrowband (#8, 10). Since, the end quality ratings were similar to each other independently whether three or four switches within of a call occurred. This is yet another argument that frequent and rapid changes of speech quality are not appreciated by the users.

As a result, the changeover between narrowband and wideband codecs turned out to be one of the main effects that influenced the quality perception in this study. However, until now the effect was observed in connection with network handovers only. To investigate how an isolated switch of the speech bandwidth is perceived without any cross-influence of packet loss or network handover, Test S3 was carried out. In opposite to the previous tests, Test S3 was designed in a more systematical way and additionally it was focused on a multi band system in which narrowband, wideband, and also super wideband codecs coexist. The magnifier of this test was put on the perception of switching between wideband and super wideband quality, but narrowband changeovers were included in the test as well to provide a link to the previous tests.

Test S3 has proven the difference between constant narrowband, wideband, and super wideband speech quality (cf. Fig. 4.4). If the super wideband speech would be

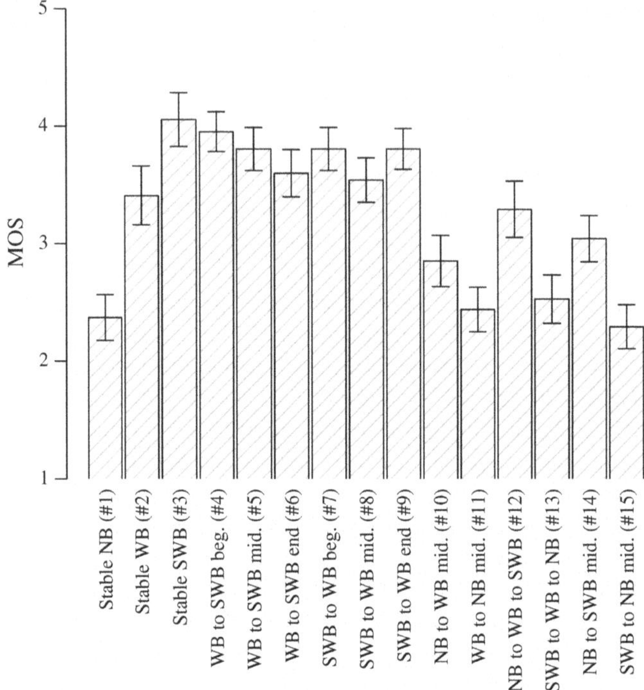

Fig. 4.4 User judgements of call quality in Test S3a. MOS_a, and 95 % CI

considered as the maximal possible quality rating of this test, the narrowband quality corresponded to 58.51 % and the wideband quality to 84.19 % of the maximal value. Apparently, the perceived distance between narrowband and wideband was larger than that between wideband and super wideband quality. To this end, the highest quality reference in all of the conducted call quality tests was rated similarly, which confirms the results of other studies that users tend to a compressed use of the MOS scale [7], even if substantially higher signal quality is provided.

In general, the ratings of conditions that were affected by switching between wideband and super wideband revealed some perceptual trends. The end quality of a call was always judged between that of pure wideband and that of pure super wideband. Moreover, if the bandwidth was increased from wideband to super wideband, the earlier in a call the change happened, the more beneficial the switch was, resulting in a substantial quality improvement only when the codec was switched at the beginning of a call (#4, 5, 6). The perception of a codec changeover in the opposite direction, i.e. from super wideband to wideband, was not defined by any systematic trends (#7, 8, 9). Whether it was switched at the beginning or at the end of a call, the end quality was rated similarly to each other and closer to that of pure super wideband rather than of wideband. This is a less critical effect as compared to a switch when narrowband was involved, because in this case decreasing the speech bandwidth in Test S3a degraded the user experience to a level that was close to pure narrowband quality (#11, 13). This indicates that the larger the relative difference between the encoding bandwidth of two codecs in a switch is, the stronger is the impact of a changeover. As a confirmation, a switch from wideband to narrowband was perceived slightly better than a direct switch from super-wideband to narrowband. In the opposite direction, if the bandwidth was increased from narrowband to either wideband or to super wideband in the middle of a call (# 10, 14), the overall quality gain was slightly higher in the super-wideband case, and the end quality experience was always substantially improved, as compared to pure narrowband. As a result, it seems that more time is required to take perceptual profits of an instant quality improvement, but rapid quality degradations are perceived immediately [8]. However, the overall experience of a switch between narrowband and super-wideband was always improved by an intermediate switch to wideband as a transition state (#12, 13). This is yet another argument that graceful fading between substantially different levels of quality is expected to improve the user experience when codecs must be switched.

The above presented results of user perception of quality during emulated calls helped to derive some guidelines for mobility management and for service quality adaptation in heterogeneous networks. Apparently, a handover between networks that require either narrowband or wideband speech codecs can be an efficient tool for quality improvement only in case packet loss impairs a connection substantially. However, if degradations due to packet loss and speech bandwidth switching can be foreseen, and a stable narrowband network is constantly available, it is preferred by the users to experience lower, but stable narrowband speech quality. Also multiple handovers between wideband and narrowband should be avoided, as rapid changes of speech bandwidth are experienced worse than pure narrowband. In general, rapid quality degradations due to switching to a codec that supports a narrower bandwidth of the speech signal are perceived immediately in a call, but users require more time to take profits of improved quality. Thus, only when the speech bandwidth is increased early in a call a substantial improvement of the overall call quality is possible, and a changeover at the end of a call may even degrade the user experience. In any case, the perceived effect is related to the difference between the bandwidths provided by the codecs that were involved in a switch, being the most critical if a connection is terminated in narrowband. In most of the cases delivery of low, but stable quality, improves not only the user experience, but also the scalability of a system as the number of quality management operations can be reduced.

4.2.2 Quality of Short Samples

As already shown in the previous section, the experience of the overall call quality is related to the stability of the transmission. Therefore, to evaluate the short-term perception of time-varying quality in heterogeneous wireless networks, a series of sample tests, the "b" tests, were carried out. The results of Test S1b (cf. Fig. 4.5) confirmed that the test participants benefited from the wideband speech coding only if the transmission channel was not affected by packet loss. If the link-quality was degraded by 10 % of packet loss, the quality judgements of the narrowband and wideband connections were similar. If the transmission channel was affected by 20 % of packet loss (that was the highest value during the test), the quality ratings of narrowband speech were higher than in the wideband case. As already mentioned in the previous section, this effect is related to a more robust packet loss concealment that was integrated in the narrowband codec [5]. Therefore, this finding also confirms the previous results that packet loss occludes the perceptual gain of a wider transmission bandwidth, and when degradations occur a robust packet loss concealment rather than the bandwidth of the speech signal is essential to provide high Quality of Experience.

Furthermore, the evaluation of codec changeover conditions in Test S1b showed that any switch between wideband and narrowband was always rated substantially lower than pure narrowband transmission. The switching direction was not of much relevance for the short-term perception of quality, although the user experience was slightly higher if wideband was provided at the end of a sample. As a result, it was proven that a rapid change between wideband and narrowband speech has a negative impact on the short-term perception of quality, and that instant quality degradation is immediately perceived, but users require more time to recover after a switching event and to take profits of wideband speech. This observations also explains the judgements of call quality from the previous tests when codec changeover was included.

If one codec was used for the entire transmission, the network handover itself turned out to have only a minor impact on the Quality of Experience. A handover from WiFi to HSDPA slightly degraded, a handover in the opposite direction slightly improved the quality perception as compared to constant transmission in the initial network. Apparently, switching to a higher or lower quality network is perceived along with the switching direction, even though the perceived effect is marginal. This, however, makes the network handover a useful tool to cope with packet loss when the same speech codec can be used in the new network. If the initial connection quality was affected by packet loss of 10 % that led to a similar quality rating of both, narrowband and wideband, a handover to a loss-free network always improved the overall experience. Even though the end quality was rated similarly in both cases, the improvement was considerable if the wideband codec was used only, because substantially higher quality was experience after the handover. To this end, this result also indicates that a drastic change of speech quality while preserving the encoding bandwidth does not degrade the user experience as a codec changeover does. As a result, the power of quality adaptation by network handover can be exploited if a substantial quality gain is expected after the switch. This is the case when strong connection degradations are experienced or if the speech codec is not robust against packet loss. Otherwise, the impact of a network handover on the user perception is marginal.

If both, network and codec switch, were combined within a sample, the user judgements were similar to that of conditions that were affected by codec changeover only. Apparently, the phenomenon of switching between wideband and narrowband

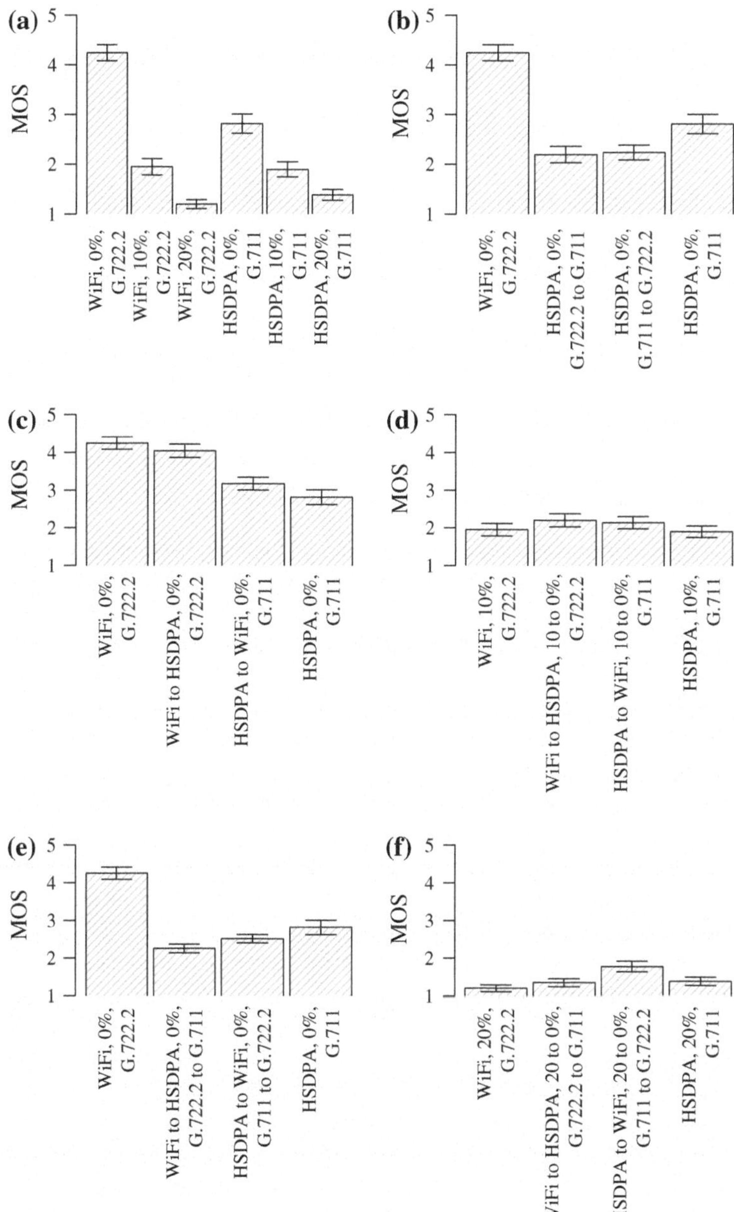

Fig. 4.5 User judgements of selected condition groups in Test S1b. MOS_a, and 95 % CI. **a** Perception of stable transmission. **b** Perception of codec changeovers. **c** Perception of network handovers. **d** Perception of network handovers, when initial connection is affected by 10 % of packet loss. **e** Perception of a combined network and codec switch. **f** Perception of a combined network and codec switch when a connection is affected by 20 % of packet loss

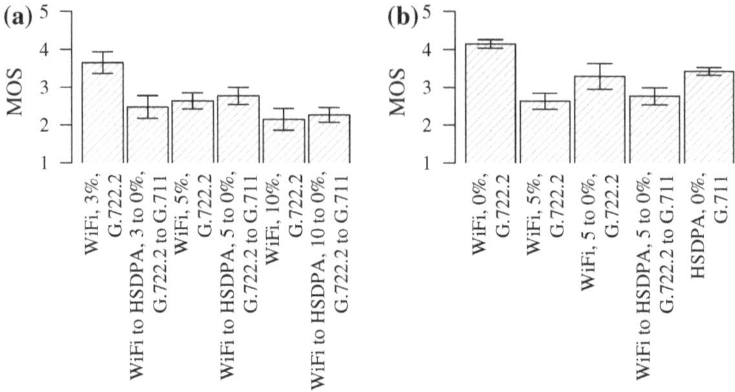

Fig. 4.6 User judgements of selected conditions groups in Test S2b. MOS_a, and 95 % CI. **a** Perception of network and codec switching when packet loss occurs. **b** Perception of packet loss discontinuity

speech dominated the user experience in this case. However, in contrast to an isolated codec changeover, if it was switched to wideband along with a handover to WiFi, the end quality was judged substantially higher than in the opposite switching case; this effect might be related to an additional gain of quality in a handover to a higher-quality network. In any case, the end quality of the combined conditions was rated below that of pure narrowband which confirms that when no other degradations are experienced no benefits of switching the network along with the codec can be achieved. However, if a connection was affected by 20 % of packet loss, a handover to a loss-free network along with a codec changeover always improved the end quality perception. This observation confirms that if the link-quality exhibits a certain degree of packet loss, the negative effect of a codec changeover can be occluded, and a quality gain through adaptation can be achieved. However, the improvement was only substantial if pure wideband was provided at the end of a sample.

A more detailed analysis of the occlusion by packet loss of the effect of codec changeover was possible based on the results of Test S2b (cf. Fig. 4.6). If the initial sample quality was high and the connection impairment was lower than 5 % of packet loss, then the negative impact of network and codec switching was remarkably high. But if the initial sample quality was degraded due to 5 % of packet loss or more, then a switch to a loss-free narrowband network improved the user experience. However, the improvement was never substantial and it was approximately constant, independently of the degree of the initial degradation. This confirms the observation that until a certain threshold of packet loss in wideband is not exceeded, a network handover along with codec changeover is not recommended. As an alternative comparison, 5 % of packet loss was disabled after the middle of a sample in the same network, which approx. corresponds to a network handover while keeping the wideband codec. The resulting quality gain was substantial, which additionally confirms previous observations that a network handover without switching of the speech bandwidth is a more suitable way of quality adaptation, and that users of wideband telephony

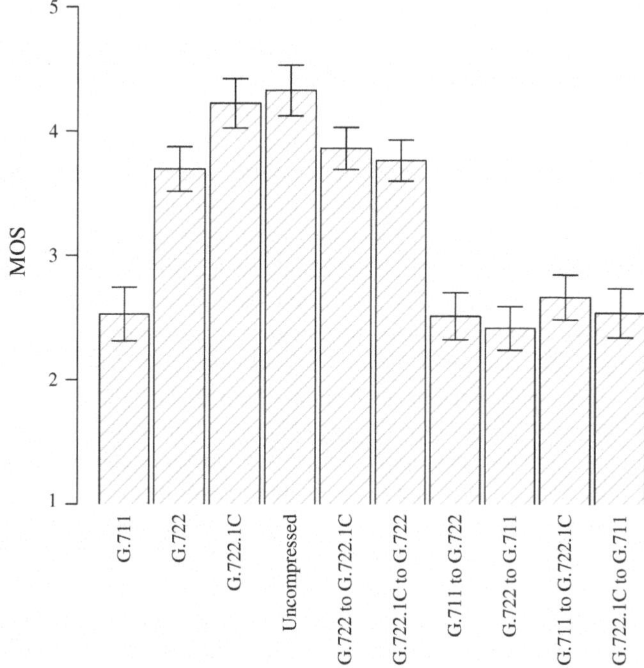

Fig. 4.7 User judgements of test conditions in Test S3b. MOS_a, and 95 % CI

benefit more of the speech quality adaptation when switching between wideband and narrowband speech is avoided. Even though the end quality in this situation was once again comparable to that of pure narrowband transmission, a substantial improvement of user experience is expected during longer calls, when more time is available to take profits of the improvement and the negative impact of a codec changeover can be avoided.

Further analysis of user perception in a multi band system based on the results from Test S3b confirmed the substantial differences in the perception of narrowband, wideband, and of super wideband speech (cf. Fig. 4.7). In general, the super wideband quality was perceived slightly worse than that of the uncompressed reference. Therefore, to provide an analysis that is comparable with the previously presented study of call quality, in the following, super wideband quality is considered as the maximal judgement of Test S3b. So, wideband quality can be expressed as 87.44 %, and narrowband as 59.71 % of the maximal value. This confirms that the different types of stable quality are judged similarly in context of short-sample and call quality tests, and that the difference between narrowband and wideband was perceived more clearly than that between wideband and super wideband also during short transmission samples.

Moreover, similarly to the corresponding Test S3a, Test S3b also confirmed the compressed use of the MOS scale by the test participants. The highest quality reference was judged similarly as in the previous tests, although a substantially wider bandwidth of the speech signal was provided in Test S3b. To this end, the judgement of the reference narrowband quality in this experiment was lower than in the other short-sample experiments due to the lack of conditions that were affected by packet loss in Test S3. As a result, the relationship between the ratings of stable narrowband and of switching conditions was different than in previous tests. If a narrowband codec was involved in a changeover, the end quality was always rated similarly to that of pure narrowband , and not below it like in the previous tests. Whether the narrowband codec was replaced with a wideband or a super wideband one the perceived quality was similar, although it was slightly higher when the super wideband codec was involved. In turn, the end quality was rated close to pure wideband, if the codec was switched between wideband and super wideband. As a result, in any case of codec switching the end quality of a sample was always rated close to that provided by the lower-quality codec that was involved in a switch. This finding also confirms previous observations that rapid quality degradation is immediately perceived, but more time is required to take perceptual profits of an instant quality improvement.

The above presented evaluation was conducted to explore the short-term perception of major transmission phenomena that are expected in heterogeneous wireless networks. The analysis explained some of the results from the previous call quality tests, as well as provided some more guidelines of speech quality adaptation that are particularly relevant late in a call, when the short-time perception is stimulated most. The study confirmed that using wider bandwidth of the speech signal improves the user experience only if the transmission is not affected by packet loss. Otherwise, robust packet loss concealment is essential to provide high Quality of Experience. Also a handover to a loss-free network without speech codec switching turned out to be a more powerful tool for quality adaptation than the combined approach. However, a considerable improvement of the overall quality was only possible if a substantial degradation of the network link was experienced. Otherwise, the impact of the network handover on the user perception was marginal. Moreover, switching between two speech codecs that support different signal bandwidth was in short term experienced similarly to each other, independently from the switching direction. This confirms previous findings from the call quality tests that degradation of quality is perceived quicker than quality improvement. The latter requires considerably more time to generate perceptual profits [8]. In a system that supports super-wideband speech quality and is not affected by packet loss, any switch between two codecs was rated close to the constant quality provided by the lower-quality codec. In a wideband-limited system that suffers from packet loss, any switch between wideband and narrowband degraded the Quality of Experience substantially below that of pure narrowband. Therefore, users of wireless networks benefit more of the adaptation of speech quality when switching between wideband and narrowband is avoided, or if a considerable amount of time remains to take the profit of such a switch. Only if substantial link-quality degradation occur that was quantified as higher than 5 % of packet loss in this work, the negative effect of wideband-narrowband codec changeover is occluded and a limited quality gain can be achieved.

4.2.3 Quality of Conversational Calls

The results presented until now in this chapter were focused on quality perception in a listening situation only. To overcome this limitation and to create a link to user experience of speech quality during a real conversation, Test S4 was carried out. Despite a more realistic value of this test, the performed test calls were up to 180 s long, which required additional pre-filtering of the collected results. Too short conversations that did not allow to present complete quality profiles, i.e. profiles in which the duration of the final quality block after the last switch was shorter than 6 s, were removed; the selection of the interval of 6 s was determined by the average duration of a test sample in the previous tests. The remaining results were divided in groups depending on which of the testbed nodes, the Mobile or Correspondent Node, they were collected. This grouping was performed due to the asymmetrical link-layer characteristics of the HSDPA network. The resources of the downlink are higher than those of the uplink [9]. Accordingly, an asymmetric impact of a network handover on user perception was expected.

Due to this data preparation method, the final data set was reduced to a maximum of 13, but not less than 6 quality judgements per condition (cf. Table A.13). A statistical screening of the test subjects was skipped in this case, as low reliability of the data was expected. However, some relevant results could still be extracted and the limitations of the test were accepted in lieu of a more realistic test protocol. Moreover, this test is relevant for future research on the perception of time-varying quality in a conversational situation.

Furthermore, an additional emphasis is required to discuss the advantage and disadvantage of an error that was discovered in the experimental setup after the test has been already carried out. The packet loss generation was incorrectly configured, and if applied, only the transmission from the Correspondent Node to the Mobile Node was degraded. Nevertheless, this issue enabled to investigate even more aspects of quality perception. So, quality judgements that were obtained on the Mobile Node were used to evaluate mobile user perception of codec changeover and network handover in the context of a transmission that might be affected by packet loss. This setup was compliant with that of the previously conducted listening-only tests. The results that were collected on the Correspondent Node, which is a different measurement point as compared to the previous tests, addressed corresponding user perception of plain codec changeover and network handover in a context of loss-free transmission.

The test results (cf. Figs. 4.8 and 4.9) confirmed that also in a conversational situation the wideband quality was experienced better than the narrowband quality and the perceived difference was similar to that during the listening tests. If a wideband connection of mobile users was degraded in the middle by a period either of narrowband or of packet loss, the end quality experience was only degraded in case the speech bandwidth was switched. While 30 s of 5 % of packet loss in the middle of a call resulted in a quality judgement that was comparable with that of pure wideband, a temporary switch to narrowband resulted in Quality of Experience that was

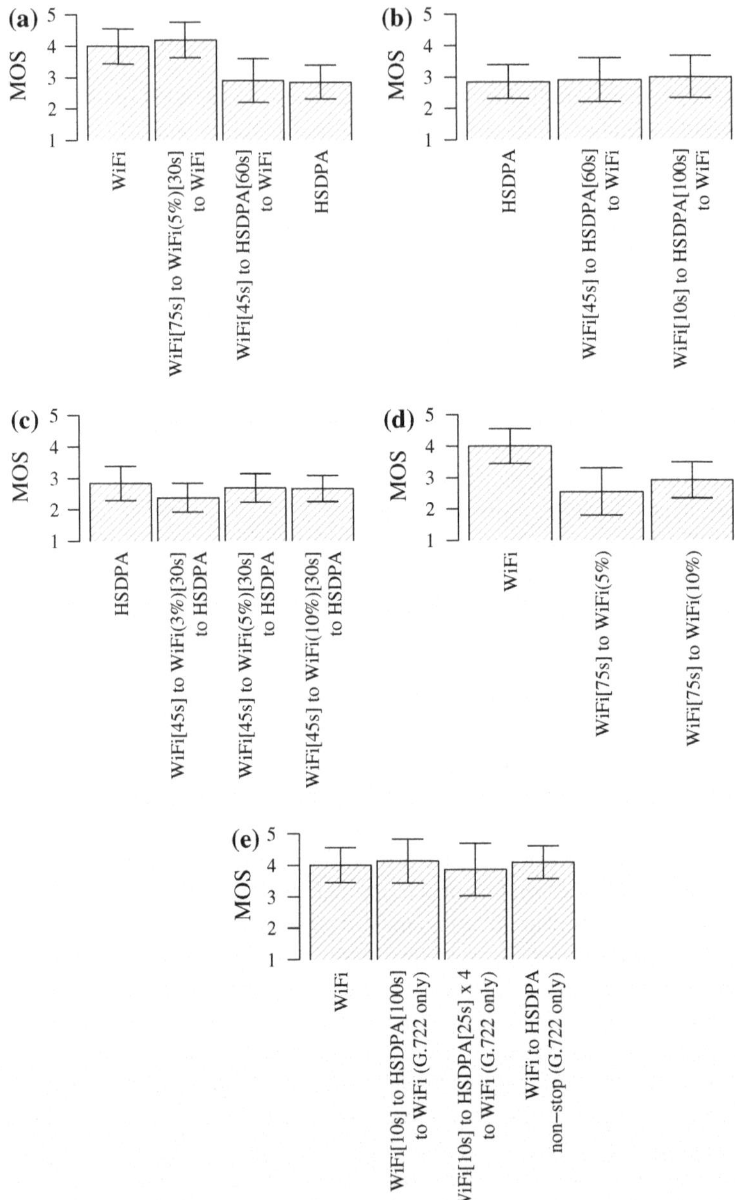

Fig. 4.8 User judgements of selected groups of calls on the Mobile Node in Test S4. MOS_a, and 95 % CI. **a** Perception of various types of degradation in the middle of a call. **b** Perception of narrowband quality in the middle of a wideband call. **c** Perception of packet loss in the middle of a call before a handover. **d** Perception of packet loss at the end of a call. **e** Perception of a network handover

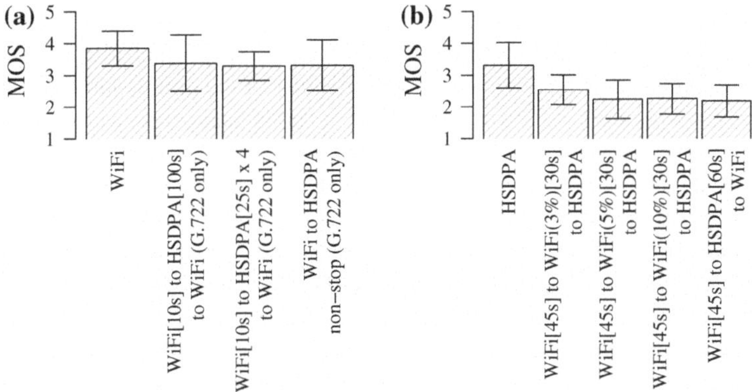

Fig. 4.9 User judgements of selected calls on the Correspondent Node in Test S4. MOS_a, and 95 % CI. **a** Perception of network and codec switching. **b** Perception of narrowband quality at the end of a call

close to that of to pure narrowband transmission. Moreover, in the case a period of narrowband quality was experienced in a call the user perception was not correlated with the duration of the low quality phase. Apparently, users are less sensitive to transmission degradations that occur not directly at the end of a real call, and in such a case only strong impairments are perceived.

A similar end quality impression was gained if the switch to narrowband was performed at the end of a call. Also in this case, the end quality was rated comparably to that of pure narrowband, irrespectively of any degradations that occurred in wideband prior to the switch. The result is consistent with the results from the listening-only tests. Once the codec was switched from wideband to narrowband, the overall quality experience was biased and the end rating was comparable to that of pure narrowband transmission; a switch in the opposite direction could not be addressed in this test due to the lack of relevant quality profiles.

Apparently, in a conversational situation packet loss influenced the overall quality experience at the end of a call only. However, the quality of conditions that were affected by packet loss was not always judged according to logical expectations. Increasing packet loss rate was not always perceived as a proportional degradation of the speech signal. Once again, this behaviour might be related to the fact that users are less sensitive to decent quality changes in a conversational situation, or that switching the talk turns between the test participants led to a longer exposure of the corresponding user to packet loss. The latter, however, cannot be proven due to the aforementioned setup error.

A minor impact on the perception of the mobile users had also the network handover. If one speech codec was used for the entire transmission, the end quality was always perceived close to that of pure wideband, independently from the number of network handovers that were performed during a call. However, even though the network handover was rarely perceived on the Mobile Node, this operation created a considerable degradation of quality that was perceived on the Correspondent Node.

Bursty packet loss was experienced during the first handover to HSDPA due to the connection initialization and resource allocation process [9], which also confirms the initial expectations that the asymmetric specification of the uplink and downlink affects the user experience. Moreover, the effect of bursty packet loss was not occluded by the changeover between wideband and narrowband codecs. Instead it amplified the perceived degradation. As a result, when a network handover is scheduled, more care of the perception of the corresponding rather than mobile users must be taken.

The analysis of user experience in a conversational situation that was presented in this section confirmed some of the results already discovered in the listening-only tests, but also helped to reveal some new aspect that are relevant for management of speech quality during real conversations in wireless networks. Pure wideband quality is experienced substantially better than narrowband also during real conversations, and the effect of codec changeover between them had the strongest impact on the user perception of all addressed degradation types. In fact, once it was switched from wideband to narrowband, the quality degradation is well remembered by the users and the end quality experience is comparable with that of narrowband transmission. Therefore, there is no perceptual gain of hopping between wideband and narrowband quality, if narrowband can be provided during the entire conversation. Other degradations, such as packet loss, impact the end quality experience only then substantially if they happen late in a call, which is different from the listening-only tests. Apparently, the size of the "recency effect" is lower in the conversational context from a listening-only situation. Moreover, in contrast to the listening-only tests, network handover turned out to be causing a considerable quality degradation on the Correspondent Node, even if it was rarely perceived on the Mobile Node. This also reveals that the Quality of Experience of the corresponding users must be considered during mobility management.

The experiment also showed that it is a complex task to conduct a conversational test under controlled conditions, when time-varying transmission conditions are addressed. A lot of test data had to be rejected to select test calls of comparable duration before the actual analysis of the user perception could be started. This problem occurred even if conversation scenarios that imply a rigid time structure, such as the ones of short conversation tests, were used. Therefore, to overcome this limitation, a careful design of quality profiles that include less dynamics is recommended. Otherwise, more test participants must be invited to such a test and more test results must be rejected to obtain a reliable data set.

4.3 Instrumental Quality Prediction

Subjective experiments are the most accurate way of speech quality assessment, but real VoIP deployments require reliable tools that help to instrumentally estimate the quality of the provided service. However, the currently available quality prediction models have neither been designed for estimation of time-varying quality, nor have they been fully validated for use in time-variant wireless environments. Therefore, the accuracy of the common sample and call quality prediction models is evaluated in this section. The analysis is based on a comparison of the user judgements that were collected during the previous tests with quality estimates provided by the models. The quality metrics are compared per condition on the 5-point MOS scale without any intermediate data mapping to guarantee relevance of the study for real deployments. Pearson correlation and Root Mean Square Error (RMSE) are used for the evaluation and the significance of the estimation errors is compared between the models according to the methodology proposed in [10], as RMSE is considered as a decisive accuracy factor in this study. The reasons of estimation errors are discussed, and some extensions of parametric modelling are proposed. The analysis in this section is based merely on the results of Test S1 and S3 for modeling of wideband and super-wideband quality, where Test S1 is selected as the most reliable data set in the domain of wideband speech (cf. Table A.1). The results from Test S2 are used solely in Sect. 4.3.2 as a supportive data set to develop a model extension.

4.3.1 Accuracy of Sample Quality Models

Quality of speech samples that are typically few seconds long can be measured by using tools that are recommended by ITU-T. The quality prediction models differ in their purpose, the range of degradations they can cope with, and the functional principle (cf. Sect. 2.2.2). However, none of them was initially developed for the assessment of sample quality in heterogeneous wireless networks. This lack of knowledge is addressed in this section, where the quality prediction that is provided by the following three models is evaluated:

- ITU-T Rec. P.863 (aka "POLQA"),
- ITU-T Rec. P.862.2 (aka "wideband PESQ"),
- ITU-T Rec. G.107.1 (aka "E-model").

The models were applied to estimate the quality of samples that were used during Test S1b and S3b. In the first case, the predictive accuracy of all three models was evaluated. However, only a subset of two models, a modified version of the E-model and POLQA, was used for quality estimation in the super-wideband context of Test S3b. The use of different sets of models arise out of the limited range of quality that is supported by most of the models. Although all of them are recommended for the assessment of wideband speech quality, only POLQA is a model that is standardized for measurement of super-wideband speech quality. The E-model as such is limited to wideband as well, but an effort towards an extension has already been made in [11]. Independently from the cited literature, this model was experimentally extended in this work for a data set that does not include any further degradations than bandwidth switching.

The proposed extension was realized by applying a linear interpolation of quality scores on the R-scale. The user judgements that were collected during Test S3b were first transformed on the wideband R-scale using the formula from [12], and then a linear interpolation was performed using the fact that the reference rating of quality provided by the wideband G.722 and by the narrowband G.711 codec on the wideband scale are known. As a result, a reference quality of $R_0 = 147.3$ was calculated for the uncompressed signal, and accordingly the impairment factors of $Ie_{G.722.1C} = 5.0$, $Ie_{G.722} = 23.3$, and $Ie_{G.711} = 54.1$ were obtained for the individual codecs. This interpolated gain of 18.3 points on the R-scale is lower than that of 40 R-points presented in [11]. However, it is emphasised that both approaches were based on data sets that included different degradation types. In general, the methodology that was applied in this work might be far too simplistic, but the results are used for a general comparison only.

The initial design of the E-model also does not enable to estimate time-varying quality. Therefore, in case a sample was composed of two blocks of different quality, a mean quality estimate on the MOS scale was computed. The mean value allows to take different quality levels into account, but also provides an unchanged score if the two blocks are of equal quality. Even if this approach might be far to simplistic, it is used for the initial benchmark only.

The collected results revealed (cf. Table 4.1) that in the wideband context of Test S1b, the quality measurements that were obtained with POLQA were the most accurate of this benchmark, which is represented by the highest correlation of the quality estimates with the user opinion, as well as by the lowest prediction error. The second most accurate result in terms of both, Pearson correlation and RMSE, was achieved when WB-PESQ was applied to predict the wideband quality. Use of the E-model resulted in the highest RMSE in any case, although in case of the super-wideband Test S3 the results provided by POLQA were less correlated with user perception.

The measured RMSE values were significantly different from each other in the wideband context of Test S1a and they were statistically equal in case of the super-wideband Test S3b. The second observation was apparently related to a lower number of conditions that were investigated in the second test. In any case, POLQA was the most accurate full-reference model in the presented comparison, and the quality assessment using the E-model was inferior in terms of RMSE to all of the tested full-reference models. However, it should be noted that none of the models was ever designed to be used in heterogeneous networks, and that the simple application of the E-model—by averaging two distinct MOS predictions for the different sample degradation levels—may be far too simplistic.

Table 4.1 Accuracy of sample quality estimates in Test S1b and S3b

	Test S1b		Test S3b	
Name	Pearson	RMSE	Pearson	RMSE
POLQA	0.96	0.29	0.83	0.62
WB-PESQ	0.95	0.43	NA	NA
E-model	0.74	1.33	0.91	0.88

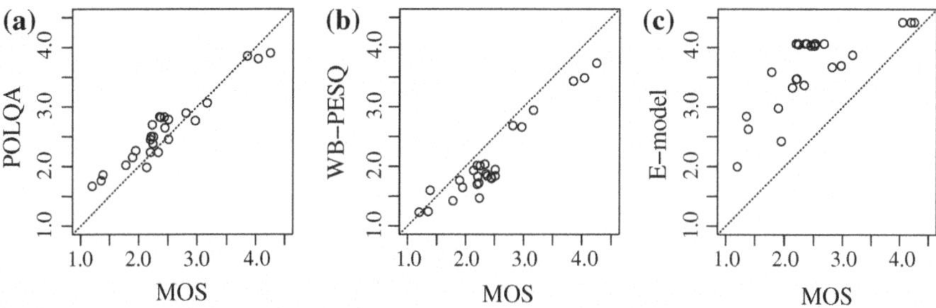

Fig. 4.10 Comparison of quality estimates and user judgements in Test S1b. **a** POLQA. **b** WB-PESQ. **c** E-model

A direct comparison of the individual quality estimates with the user judgements that were collected in Test S1b (cf. Fig. 4.10) confirmed that predictions provided by both, POLQA and WB-PESQ, were close to user opinion. However, certain differences between the models were observed. The quality estimates provided by POLQA were predominantly overestimated in the lower parts of the scale, which discloses that conditions that were affected by the highest packet loss rates of the test suffered from the most inaccurate estimation. Moreover, this model also collectively overestimated a bunch of conditions that were judged by the users in the range between approx. 2.2 and 2.5 MOS. This group included the majority of codec changeover conditions that were not affected by packet loss. The same subset of conditions was especially critical when the WB-PESQ was used. In this case, the quality was considerably underestimated, which goes along with the general tendency of this model to underestimate the quality in the entire quality range with some exceptions in the low quality area. The use of the E-model resulted in a systematical overestimation of most of the test conditions. Especially inaccurately assessed were conditions that were affected by codec switching as well as by changing packet loss.

A considerable overestimation of quality was also observed in the super-wideband context of Test S3b (cf. Fig. 4.11). Both of the evaluated models, POLQA and the E-model, significantly overestimated the quality of conditions that included narrowband speech. In particular, switching from either wideband or super-wideband to narrowband turned out to be the most critical to assess for both models, but a systematic error was observed if POLQA was used. This model estimated the quality of conditions that were terminated in narrowband significantly more positive than the quality of a switch in the opposite direction. This behaviour is contradictory to the analysis of user perception presented in the previous section, where it was shown that a sample affected by a switch in any direction is perceived similarly. This principle, however, was well covered by POLQA if a switch between wideband and super wideband codec was performed. In this case, also the E-model provided more accurate results, although the inaccuracy was also considerable. It is worth noting that pure narrowband quality in Test S3b was also significantly overestimated by both models, and an error of 0.73 and 0.82 MOS points was introduced in this case the E-model

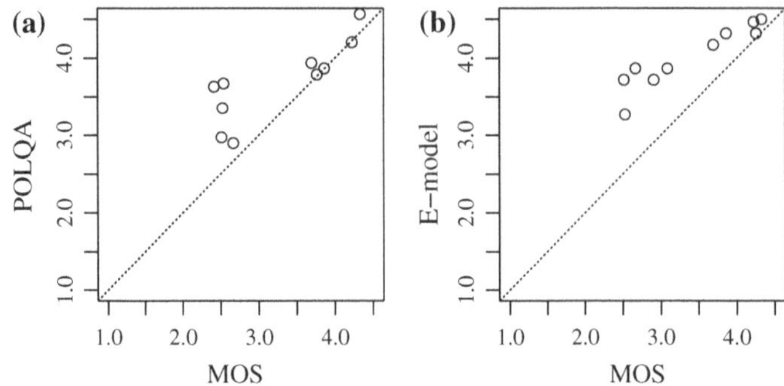

Fig. 4.11 Comparison of quality estimates and user judgements in Test S3b. **a** POLQA. **b** E-model

and POLQA respectively. This problem might be related to the aforementioned fact that no further degradations by packet loss were included in the subjective experiment and thus the quality perception of the low quality reference was shifted to the bottom of the scale, as compared to other user tests that include signal degradation by packet loss. This observation leads to a conclusion that in a multi band environment that is not affected by any further degradations, benchmarking of the coding quality is inaccurate when the above models are applied. In this particular situation, other models that are trained for this purpose should be used.

To confirm the above observations and to identify significant reasons of estimation error in Test S1b, statistical analysis was used. The Analysis of Variance (ANOVA) with significance level of 5 % was computed for each of the models separately. As the depended variable of the linear model, the estimation error per condition was used, and as the independent variables speech codec, network, and packet loss were applied. The parameter switching was modeled through additional levels of the factors, which allowed to analyse the time-varying quality. The normality of the error population was examined visually; in case of Test S3 ANOVA was not applied, because only one dependent factor was included in the test.

The obtained results (cf. Table 4.2) confirmed that the accuracy of the quality estimates provided by POLQA was significantly affected by the speech codec and packet loss, in the respective order of relevance. As observed before, switching

Table 4.2 Statistical contribution of transmission parameters to the quality prediction error in Test S1b

	POLQA		WB-PESQ		E-model	
Factor	F	p	F	p	F	p
Network	1.34	0.3088	2.64	0.07343	19.74	1.2e-05
Loss	4.56	0.0160	12.35	0.00023	3.87	0.028
Codec	6.44	0.0066	7.10	0.00454	66.77	3.7e-08

between wideband and narrowband codecs and strong packet loss degradations were the main challenges for the model. An inverse order of relevance was observed for WB-PESQ whose prediction accuracy was affected by packet loss the most and less by the speech codec. Additionally, the network slightly contributed to the estimation error, which indicates that changes of codec and packet loss due to handover imposed an additional challenge for this model. The quality prediction that was provided by the E-model was significantly affected by the speech codec, network, and packet loss, in the respective order of relevance, which is a pointer to the largest E-model error that was observed for the majority of the combined network and codec switching conditions, as well as conditions in which packet loss was disabled after a network handover.

The analysis presented in this section has shown that the use of speech quality prediction models in heterogeneous networks will be determined by the ability of a model to predict time-varying quality. In the group of the evaluated models, only two of the signal-based models, with POLQA being the most accurate, provided quality estimates that were close to user opinion. However, both of the models showed a weakness in the assessment of quality when a sample was affected by narrowband-wideband codec changeover or strong and instable packet loss degradations. The parametric E-model in the simplistic form as proposed in this section failed to assess the quality. This does not come at a surprise, as the signal-based models have far more input information on the quality of the particular speech stimuli available than a parametric model has. More parametric modelling of time-varying transmission is required to improve the accuracy of the E-model in context of heterogeneous networks, and an approach how to do this is proposed in the next subsection. It has also to be emphasised that all of the above presented results could be improved by an intermediate adjustment of the collected data. Procedures such as third-order mapping are frequently used to reduce the effect of a user test during development of quality prediction models. However, the goal of the presented analysis was to address real data as it would be measured by a regular user, without any intermediate results tuning. Therefore, the mapping was not applied.

4.3.2 Extension of ITU-T Rec. G.107.1

As already presented in the previous section, a simplistic averaging of quality estimates provided by the E-model for subsequent parts of a sample has not proven its accuracy. This limitation of time-varying quality estimation was in particular observed when the speech codec was switched. To overcome this issue and to extend the area of the E-model application, an additive codec switching impairment factor I_{esw} is proposed in this section.

The E-model is a planning tool that is designed for assessment of speech quality provided by a transmission system under stable circumstances. The main principle of the model assumes an additive relationship of quality impairments on the R-scale that is the internal scale of the model. Thus, the main formula is expressed as follows:

$$R = R_0 - I_d - I_s - I_{e,eff} + A \tag{4.1}$$

The impairment factors I_d (delay impairment), I_s (simultaneous impairment), and I_{eff} (effective equipment impairment) are used in an additive way to degrade the reference quality R_0 that is defined as 93.2 or 129 for narrowband and wideband speech quality, respectively. The advantage factor A is applied to reflect advantages that result of some system features.

In the formula, degradations that are related to speech coding are reflected in the effective equipment impairment factor $I_{e,eff}$ that is denoted for random packet loss rates as follows:

$$I_{e,eff} = I_e + (95 - I_e)\frac{P_{pl}}{P_{pl} + B_{pl}} \tag{4.2}$$

where the codec impairment factor I_e, packet loss probability P_{pl}, and the packet loss robustness factor B_{pl} are the decisive factors. The parameters must be selected according to the reference quality R_0, and an overview of recommended values that are relevant for this study is provided in Table 4.3 for wideband and in the previous section for super-wideband quality.

As already mentioned in the previous section, computation of the mean value over estimates measured for two sub-blocks of a speech sample enables to include different levels of quality in the end rating. Moreover, this calculation also provides an unchanged value if the two blocks are of equal quality. Therefore, Eq. 4.3 is the starting point to derive the new impairment factor.

$$R = \frac{R_1 + R_2}{2} \tag{4.3}$$

The equation was used to establish the initial benchmark on the R-scale by applying it to estimate the quality of samples from Test S1b and S2b; as mentioned in the introduction of this section, the E-model extension was developed based on the results of the two wideband tests that were conducted in this work [13]. Test S3b, however, is used to validate the proposed model extensions at the end of this subsection. The collected results have been normalized to reduce the modelling error and to improve the accuracy of the proposed solution. For this purpose, the linear interpolation that is recommended in ITU-T Rec. P.833.1 [14] was applied. The obtained quality scores confirmed on the R-scale (cf. Fig. 4.12) that the estimates that were computed for Test S1b were close to user opinion only in case stable transmission was addressed.

Table 4.3 Wideband impairment factors and robustness parameters of selected codecs

Codec	I_e	B_{pl}
G.722.2 (23.05 kbit/s)	1	13
G.722.2 (12.65 kbit/s)	13	13
G.711 (64 kbit/s)	35.8	25.1

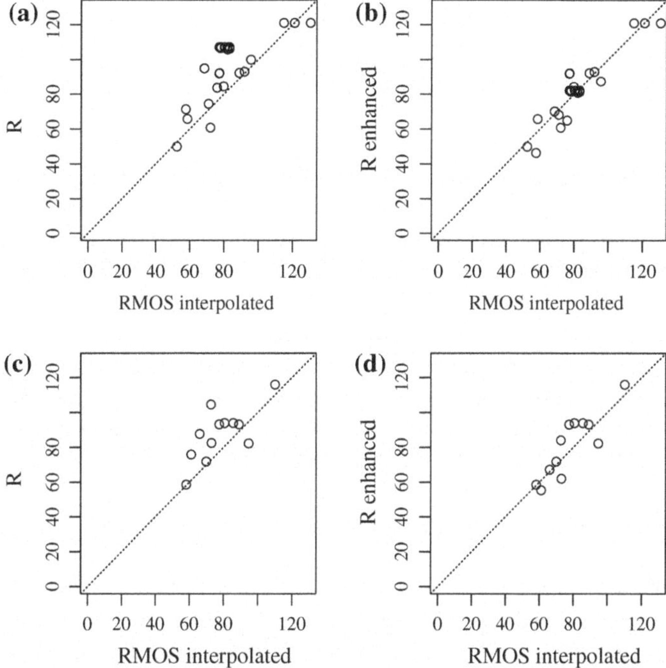

Fig. 4.12 Comparison on the R-scale of quality estimates and interpolated user judgements in Test S1b and S2b. **a** Mean estimate in Test S1b. **b** Enhanced model in Test S1b. **c** Mean estimate in Test S2b. **d** Enhanced model in Test S2b

The prediction accuracy decreased, but remained considerably accurate, when the packet loss rate was changed due to a handover to a loss-free network. However, if the speech codec was switched, the quality of the affected samples was significantly overestimated. The problem was independent of the switching direction and the measurement error varied between 29.96 and 13.59 points on the R-scale, and decreased if the effect of a codec changeover was occluded by high packet loss rates.

The main findings from Test S1b were confirmed in the data of Test S2b. The quality of codec switching conditions was the most overestimated one, and the estimation error varied between 31.93 and 9.56 points on the R-scale. In both tests, the prediction error was of comparable magnitude to the difference between quality impairments imposed by codecs involved in a switch. For illustration, a difference of 34.8 and of 22.8 points on the R-scale was measured between the impairment factors I_e of codecs that were used in in Test S1b and S2b, respectively.

Therefore, to reduce the quality prediction error that was observed during the initial benchmark, an additive codec switching impairment factor I_{esw} was introduced in the basic formula as follows:

$$R = \frac{R_1 + R_2}{2} - I_{esw} \tag{4.4}$$

where the factor itself is defined as a weighted absolute difference of the equipment impairment factors:

$$I_{esw} = a|I_{e1} - I_{e2}| \tag{4.5}$$

This definition of I_{esw} was determined by two of the above observations. First of all, the estimation error was related to the difference between impairment factors of both codecs involved in a changeover $I_{e1} - I_{e2}$. Second, the direction of a switch was not decisive, which leads to the use of the absolute difference between the equipment impairment factors; the perceptual occlusion of a changeover by packet loss was not modelled at this place due to a limited data set. For example, conditions degraded by packet loss after the switch were not included.

The coefficient a was optimized to achieve the highest prediction accuracy for the concatenated data set of both tests. For this purpose, least-square fitting was applied, and a value of $a = 0.897$ was obtained. Therefore, the final formula of the codec switching impairment factor I_{esw} is denoted as follows:

$$I_{esw} = 0.897|I_{e1} - I_{e2}| \tag{4.6}$$

The final model that consists of a combination of Eqs. 4.4 and 4.6 was ultimately used for the quality prediction of samples from the underlying data set of Test S1a and S2b and the results were confronted with those from the initial benchmark (cf. Fig. 4.12). In this way, it was confirmed that the new modelling approach enabled to reduce the overestimation of codec switching conditions in both data sets. As well, conditions that were not affected by codec switching remained near the optimum line.

The new approach of parametric quality prediction helped to more reliably estimate the quality of samples from all data sets, in particular including the super-wideband Test S3b that was not used for modelling (cf. Table 4.4). As compared to the results presented in the previous section that were interpreted on the MOS scale, the correlation of the quality estimates and user opinion could be considerably improved and the quality prediction error could be decreased by 0.53, 0.44, 0.22 for Test S3b, Test S1b, Test S2b, respectively. The strongest improvement in case of the super wideband Test S3b confirms that the choice of the relative difference between the impairment factors for modelling enables to predict the quality of samples that include a switch between wideband and super-wideband codecs, where the difference between the speech quality levels is less perceived.

Table 4.4 Accuracy of quality estimates when using the enhanced E-model in Test S1b, S2b, and S3b

E-model	Mean		Enhanced	
	Pearson	RMSE	Pearson	RMSE
Test S1b	0.74	1.33	0.90	0.89
Test S2b	0.71	0.63	0.86	0.41
Test S3b	0.91	0.88	0.95	0.35

The above presented study of quality modelling showed that even though simplistic averaging of quality estimates provided by the E-model in case of codec switching does not enable to reliably predict the quality, the inaccuracy can be reduced in both environments, wideband and super wideband, by applying the extension of the model that was proposed in this section. The presented method is a first trial to parametrically quantify the impact of speech bandwidth switching on user perception in a parametric way. However, more work is still required to improve this initial approach. For example, more transmission scenarios that include a larger set of codecs must be evaluated, and the occlusion effect of the codec changeover perception by packet loss must be modelled.

4.3.3 Accuracy of Call Quality Models

All of the models addressed until now in this section were designed for the prediction of short sample quality. This is insufficient if the quality of a transmission that is of typical phone call duration should be measured. For this purpose, so called "call quality models" were developed. Based on parametric modelling of quality scores that are measured for selected periods of a longer call, the overall Quality of Experience can be assessed [15]. In this section, the quality prediction accuracy of the following models in the context of heterogeneous wireless networks is addressed:

- model proposed in ETSI 102 506 v.1.2.1 [2] (aka "ETSI model"),
- model proposed in [15] by Weiss et al. [15] (aka "Weiss model"),
- model proposed in [16] by Rosenbluth (aka "Rosenbluth model"),
- baseline averaging of the sample quality (aka "Mean model"),

where of the actual call quality models are in focus of the following investigation, and the baseline approach of quality averaging is provided for initial comparison only.

The analysis is based on the data sets from the wideband Test S1 and the super-wideband Test S3. The quality of the 60 s long emulated calls that were introduced in the "a" tests was estimated using the above models and the results were compared to the collected user judgements. The end call quality was modelled according to the five composing samples of a quality profile using as input the MOS scores per condition that were obtained in the corresponding "b" tests.

The collected results (cf. Table 4.5) revealed that the quality estimates that were provided by all of the models were well correlated with user opinion and that only subtle differences between the compared tools were observed. The quality estimates provided by the Weiss and Rosenbluth model were equally well correlated with the user perception in the wideband context of Test S1, but slightly better results in terms of both, Pearson correlation and RMSE, were provided by the Rosenbluth model in case of the super-wideband Test S3. Moreover, both of the models were always superior to the ETSI model whose output was always affected by the largest prediction error that was even higher than that in case of pure quality averaging. The

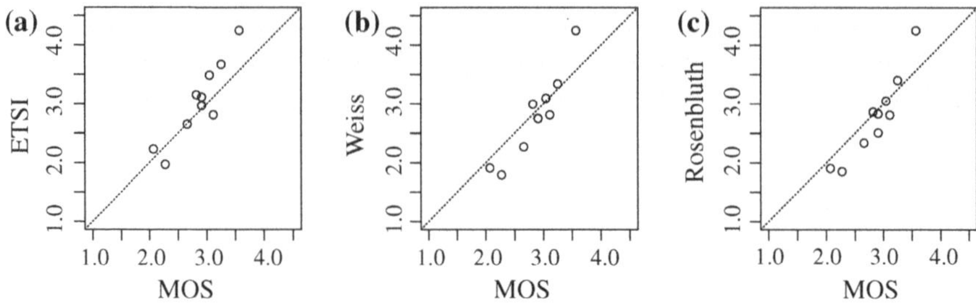

Fig. 4.13 Comparison of quality estimates and user judgements in Test S1a. **a** ETSI. **b** Weiss. **c** Rosenbluth

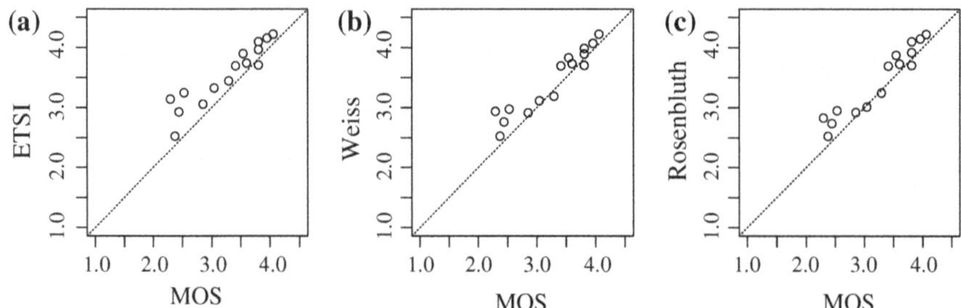

Fig. 4.14 Comparison of quality estimates and user judgements in Test S3a. **a** ETSI. **b** Weiss. **c** Rosenbluth

last, however, resulted in the lowest correlation with the mean opinion score. Finally, none of the described differences between the RMSE values that were provided by the models was significant.

Further analysis of the call quality estimates that were computed in Test S1 (cf. Fig. 4.13) showed that the reasons of prediction errors were different in case of the ETSI model, than of the two other models. The ETSI model mainly overestimated the quality in the upper part of the MOS scale, which was related to the fact that this model failed to assess the quality of calls that introduced longer periods of pure wideband. This subset includes profiles that were affected by a switch from narrowband to wideband at the beginning and in the middle, as well as by a changeover

Table 4.5 Prediction accuracy of call quality models in Test S1a and S3a

	Test S1a		Test S3a	
Name	Pearson	RMSE	Pearson	RMSE
Mean	0.91	0.34	0.90	0.39
ETSI	0.92	0.35	0.93	0.37
Weiss	0.94	0.33	0.95	0.26
Rosenbluth	0.94	0.33	0.96	0.25

from wideband to narrowband at the end of a call. Apparently, neglecting the influence of a bad sample on user perception leads to a quality overestimation. In turn, the other two models, Weiss and Rosenbluth, mainly underestimated the quality in the lower part of the quality scale. It could be identified that profiles that introduced rapidly decreasing quality, such as profiles that were affected by a loss pattern of 0, 10, 20, 10, 10 % for the composing samples, as well as the profile that included a handover from an error prone wideband to a loss free narrowband network were significantly underestimated by these two models. Apparently, the influence of a bad sample quality on the end quality experience is weighted too strongly for correct assessment of wideband transmission quality. Additionally, all three of the models suffered from considerable overestimation of the pure narrowband condition and a significant underestimation of the wideband reference sample. These observations are related to the averaging effect identified in the previous section that long exposure to a constant stimulus reduces the user sensitivity to differentiate the reference good and bad quality from the rest of the test material. The former effect is, however, additionally related to the fact that pure wideband was rated by the users in Test S1a considerably lower as compared to the other tests.

More consistent trends were explored in the super-wideband Test S3a (cf. Fig. 4.14). In this case the majority of quality estimates provided by all of the models was overestimated. Moreover, a similar set of conditions to that of the previous test with yet inverse prediction error was revealed. Conditions that were affected by decreasing quality in the middle or at the end of a call were significantly overestimated. This was the majority of quality profiles that were terminated in narrowband. The effect might be related to the aforementioned fact, that the ratings of pure narrowband quality in Test S3 "a" and "b" were lower than expected due to the fact that packet loss degradations were not included in the data set, and so a lower reference quality was not provided. As a consequence, the relative impairment of a wideband-narrowband codec switch was less negative than in the other tests.

The presented evaluation confirms that the currently existing call quality prediction models enable to achieve quality estimates close to the user opinion in both, wideband and super wideband speech transmission systems. In any case, modelling approaches that included additional weighting of bad transmission periods of a call turned out to be more accurate. However, when the transmission quality was rapidly decreasing in time due to degradations that are typical for for heterogeneous wireless networks, such as strong packet loss or speech codec changeover, the weighting of a negative event turned out to be too strong and the estimation accuracy dropped significantly. The sign of the error turned out to be related to the reference quality conditions used in the test. This finding, however, must be confirmed by more experiments. More insights are also provided in the next chapter that addresses the prediction of video calls quality.

4.3.4 Instrumental Call Quality Modeling

Ideally, prediction models that were designed for estimation of both, sample and call quality, should be used together to enable a fully instrumental approach of quality assessment. However, the relationship of the errors that would be introduced in the subsequent stages of the computation by different model types has not yet been explored for heterogeneous wireless environments. This is the subject of this section.

According to the landscape of quality models that was presented in this section, multiple model tandems are possible. However, only two of them, a fully parametric and a mixed model tandem, are considered as relevant for the praxis in the following. The fully-parametric approach is based on a joint utilization of the enhanced E-model and the Weiss model. The enhanced E-model was selected as the most accurate method of parametric assessment of short sample quality in heterogeneous networks. The Weiss model was selected as one of the best call quality models that requires lower computation complexity as its opponent, which makes this modes more valuable for real deployments. Both models are parametric models, so that no access to the speech signal is necessary, and so real-time monitoring of call quality is possible with this approach. In opposite, the mixed model tandem employs the full-reference signal-based POLQA model instead of the enhanced E-model. POLQA provided the most accurate estimates of sample quality in this chapter. Therefore, this approach is more suitable for offline quality diagnostic.

The estimation error of both approaches was evaluated in analogy to the previous section, and the analysis was based on the data from the wideband Test S1 and the super-wideband Test S3. However, instead of the user judgements from the "b" tests, the quality estimates that were measured with the respective models were used as the modelling input.

The obtained results are consistent with logical expectations (cf. Table 4.6). When the models were applied to assess the quality of calls from the wideband Test S1, different effects were observed. While the estimates obtained with the mixed model tandem were close to user perception, the output of the parametric approach was well correlated with user perception but systematically overestimated. The origin of those results is related to the high accuracy of quality estimation provided by both models that were used in the mixed tandem, and to a systematical overestimation of sample quality that was introduced by the E-model in the parametric approach (cf. Fig. 4.15). In the super wideband context of Test S3, the accuracy of both, mixed and parametric, methods was lower than in the previous test. A biased overestimation of conditions in the lower parts of the quality scale was observed (cf. Fig. 4.16), which was directly related to the incorrect assessment of narrowband signal quality by POLQA and the E-model in this test.

The study of fully instrumental assessment of call quality confirmed that careful selection of quality models in heterogeneous networks is required. If models that cannot handle certain degradation types are applied, systematical errors during call quality prediction are generated.

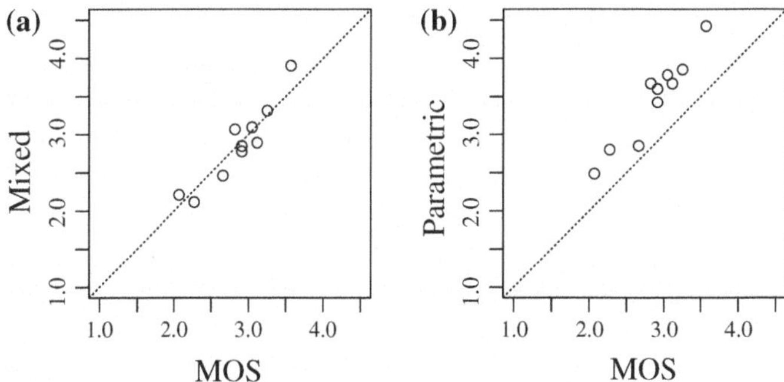

Fig. 4.15 Comparison of quality estimates by model tandems and user judgements in Test S1a. **a** Mixed. **b** Parametric

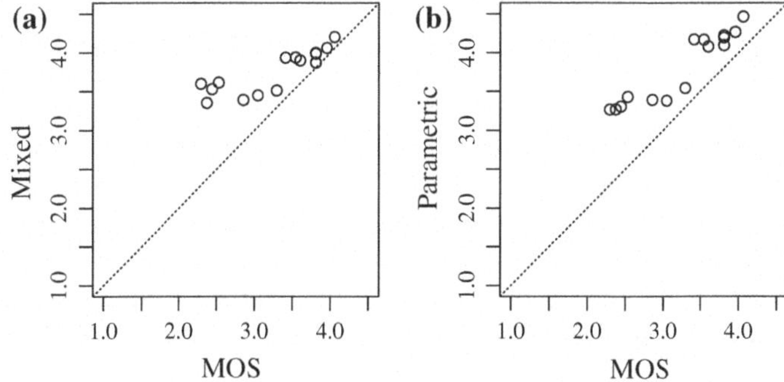

Fig. 4.16 Comparison of quality estimates by model tandems and user judgements in Test S3a. **a** Mixed. **b** Parametric

4.4 Chapter Summary

The study that was presented in this chapter helped to understand user perception of speech quality in heterogeneous wireless networks. A set of subjective experiments was performed to link user experience of quality during real calls with the perception of single transmission phenomena. A particular difficulty in the design of the tests resulted from the multitude of relevant transmission parameters and of the switching combinations that may happen in such networks. Therefore, to anchor the main

Table 4.6 Accuracy of quality estimates provided by selected model tandems in Test S1a and S3a

Name	Test S1a		Test S3a	
	Pearson	RMSE	Pearson	RMSE
Mixed	0.94	0.18	0.85	0.64
Parametric	0.96	0.63	0.94	0.61

perceptual effects an exploratory methodology was used at the beginning of this work, and then a more systematical approach was applied to understand the decent effects. Moreover, the experiments performed in this study also showed that it is a complex task to conduct a conversational test under controlled conditions, when time-varying degradations are addressed. A lot of test data had to be rejected to select test calls that match the duration of the investigated quality profiles. Therefore, a careful design of quality profiles that include limited dynamics is recommended.

The collected results of the subjective experiments confirmed that users are less analytic in their judgements during real conversations than in a listening only situation, and that the quality perception is related to the duration of the transmission and to the temporal position when a degradation or a switching event occurs. Rapid degradation of quality is experienced quicker than instant quality improvement. The latter requires considerably more time to generate perceptual profits [8]. Therefore, switching between two speech codecs of different bandwidth is experienced in both directions similarly to each other when the effect is measured directly after the switch, but the impact on the overall Quality of Experience is related to the relative position of the changeover within of a call. In fact, only when the speech bandwidth is increased early in a call a substantial improvement of the overall call quality is possible. In turn, once the codec is switched from wideband to narrowband, the quality degradation is well remembered by the users and the end quality experience is comparable with that of narrowband transmission. Therefore, there is no perceptual gain of hopping between wideband and narrowband quality, if narrowband can be provided during the entire conversation.

The perception of a codec changeover itself is relative and depends on the degradation types that are experienced during transmission. In a system that supports super-wideband speech and that is not affected by packet loss, any switch between two codecs is rated close to the constant quality provided by the lower-quality codec. In a wideband-limited system in which packet loss can be experienced, any switch between wideband and narrowband degrades the Quality of Experience substantially below that of pure narrowband. Apparently, when a very low quality reference is generated, users of wireless networks benefit more of the adaptation of speech quality when switching between wideband and narrowband is avoided, or if substantial amount of time remains to take the profit of such a switch. As a matter of fact, a changeover to a higher bandwidth at the end of a call may even degrade the user experience. Only if significant link-quality degradation occurs that was quantified as more than 5 % of packet loss in this work, the negative effect of a codec changeover between wideband and narrowband is occluded.

As a result, a handover between networks that require either narrowband or wideband speech codec can be an efficient tool for quality improvement only in case packet loss impairs a connection significantly. However, if degradations due to packet loss and speech bandwidth switch can be foreseen, and a stable narrowband network is constantly available, it is preferred by the users to experience lower, but stable narrowband speech quality. A more powerful tool for quality adaptation is pure handover to a loss-free network (without switching of the speech codec). However, a considerable improvement of the overall quality is only possible if a significant degradation of the network link was experienced. Otherwise, the impact of the handover on the perception of the mobile users is marginal. However, it has to be emphasised that if an asymmetrical connection is provided, such as a HSDPA link, considerable quality degradations during a network handover may be perceived by the corresponding users.

Packet loss as such degrades the quality perception only if its impact could be remembered by the users. Therefore, packet loss was the primary degradation during listening-only tests, but in a real call, packet loss degraded the user experience only if it affected the quality at the end of a call. Degradations in the middle of a call have not influenced the overall experience of a call in the performed conversation test. The bandwidth of the speech signal as such influences the Quality of Experience only if transmission is not affected by packet loss. Otherwise, robust packet loss concealment is essential to provide high Quality of Experience.

Even though the above analysis provided extensive insights into perception of speech quality in heterogeneous wireless networks, a set of new research questions that are necessary to further optimize the quality management arises. First of all, it is necessary to understand whether smoothening of the quality transition when the provided signal bandwidth is switched can help to substantially improve the user experience of a codec changeover, although researchers have already proven that a gradual transition that is 2.5 s long from narrowband to wideband is perceived nearly the same as an abrupt switch [6]. Furthermore, it has to be analysed whether a switch between codecs that operate at the same bandwidth of the speech signal but are differently robust against transmission impairments is a valuable tool to adapt the quality when packet loss occurs.

Nevertheless, based on the presented analysis, two groups of policies can be derived for active or proactive management of speech quality. While the first method is a direct response to unexpectedly occurring transmission circumstances, the second one enables to optimize the service quality when more information is available.

As a direct response to bad network conditions, the following policies should be applied to improve the user experience during quality adaptation:

- Hopping between different speech bandwidths should be avoided at any time.
- If the packet loss rate is lower than 5 %, only a switch to a network that supports the same speech codec is recommended.
- If the loss is higher than above, any network handover that helps to maintain the connectivity should be performed.

If more information is available for the management of service quality, the following proactive routines are recommended:

- If unimpaired delivery of the speech signal is not possible, a rapid degradation of quality is expected, and a robust wide-range network is available, the robust link should be utilized permanently.
- If a high quality connection is available in a system where packet loss is experienced, a switch to a higher quality codec is recommended only early in a call, when at least 30 s remains to take perceptual profits of the quality improvement.

- If a high quality connection is available in a system that never introduces any packet losses, a switch to a higher quality codec is recommended any time.

Once the perceptual effects have been quantified, the quality prediction models like the ones recommended by ITU-T could be validated for use in wireless environments. In the group of the evaluated sample quality models (POLQA, WB-PESQ, and the E-model) only the two signal-based models, POLQA and WB-PESQ, with POLQA being the most accurate one, provided quality estimates that were close to user opinion. However, both of them showed a weakness in the assessment of conditions that were affected by a changeover between narrowband and wideband or by strong and instable packet loss degradations. Simplistic averaging of quality estimates provided by the E-model did not enable to reliably predict the quality, but the inaccuracy could be significantly reduced by applying the extension of the model that was proposed in this chapter. However, the proposed method is a first try to parametrically quantify the impact of speech bandwidth switching on user perception and more work is still required. It also has to be emphasised that in a test where all three bandwidths, narrowband, wideband, and super-wideband were provided and packet loss was not experienced, a biased overestimation of narrowband quality was observed. Apparently, the strict training of sample quality models based on data sets that include transmission artifacts that are typical in IP networks does not allow to use these models for benchmarking of pure coding quality in multi band systems. In turn, the call quality as such could be accurately predicted with all of the available call quality models, although methods that included additional weighting of bad transmission periods during a call are more accurate. However, the weighting turned out to be too strong when transmission quality is rapidly decreasing in a call, which might be a critical effect for prediction of speech quality in heterogeneous wireless networks, where sudden changes of quality are experienced.

References

1. ITU-T Rec. P.800 (1996) Methods of subjective determination of transmission quality. In: International telecommunication union, Geneve, Switzerland
2. ETSI 102 506 v. 1.2.1 (2006) Speech processing, transmission and quality aspects (STQ); estimating speech quality per call. In: European telecommunications standard institute, Sophia Antipolis, France
3. ITU-T Rec. G.113 (2009) Transmission impairments due to speech processing. In: International telecommunication union, Geneve, Switzerland
4. Möller S (2000) Assessment and prediction of speech quality in telecommunications. Kluwer Academic Publishers, Boston
5. ITU-T Rec. G.113-Amendment 1(2006) New appendix IV provisional planning values for the wideband equipment impairment factor ie, wb. In: International telecommunication union, Geneve, Switzerland
6. Voran SD (2010) Subjective ratings of instantaneous and gradual transitions from narrowband to wideband active speech. In: Proceedings of IEEE international conference on acoustics, speech and signal processing (ICASSP). Dallas, USA, pp 4674–4677
7. Möller S (2009) Towards a universal scale for perceptual value. ITU-T Contr. COM 12–81, Geneve, Switzerland

8. Gros L, Chateau N (2001) Instantaneous and overall judgements for time-varying speech quality: assessments and relationships. Acta Acustica United Acustica 87(3):367–377
9. Love R, Ghosh A, Xiao W, Ratasuk R (2004) Performance of 3GPP high speed downlink packet access (HSDPA). In: Proceedings of IEEE vehicular technology conference (VTC), vol 5. Los Angeles, CA, USA, pp 3359–3363
10. VQEG MM-FR1 (2008) Final report from the video quality experts group on the validation of objective models of multimedia quality assessment, Phase I
11. Wältermann M, Tucker I, Raake A, Möller S (2010) Extension of the E-model towards superwideband speech quality. In: Proceedings of IEEE international conference on acoustics, speech and signal processing (ICASSP). Dallas, USA, pp 4654–4657
12. ITU-T Rec. G.107 (2011) The E-model, a computational model for use in transmission planning. In: International telecommunication union, Geneve, Switzerland
13. Lewcio B, Wältermann M, Möller S, Vidales P (2009) E-model supported switching between narrowband and wideband speech quality. In: Proceedings of IEEE international workshop on quality of multimedia experience (QoMEX). San Diego, CA, USA, pp 98–103
14. ITU-T Rec. P.833.1 (2008) Methodology for the derivation of equipment impairment factors447 from subjective listening-only tests for wideband speech codecs. In: International telecommuni448cation union, Geneve
15. Weiss B, Möller S, Raake A, Berger J, Ullmann R (2009) Modeling call quality for time-varying transmission characteristics using simulated conversational structures. Acta Acustica United Acustica 95(6):1140–1151
16. Rosenbluth J (1998) Testing the quality of connections having time varying impairments. ITU-T Contr. D.64, Geneve, Switzerland
17. Lewcio B, Möller S, Vidales P, Wältermann M, Kirschnick N (2008) Methods for multimedia service adaptation in next generation networks. In: Proceedings of the ETSI workshop on effects of transmission performance on multimedia QoS. Prague, Czech Republik
18. Möller S, Wältermann M, Lewcio B, Kirschnick N, Vidales P (2009) Speech quality while roaming in next generation networks. In: Proceedings of IEEE international conference on communications (ICC). Dresden, Germany, pp 1–5
19. Möller S, Wältermann M, Lewcio B, Kirschnick N, Vidales P (2008) Handovers between narrowband and wideband speech transmission. In: Proceedings of sprachkommunikation 2008, Vorträge der ITG-Fachtagung, ITG-Fachbericht, Aachen, Germany, 2008. VDE-Verlag

Chapter 5
Experience of Video Telephony

Similarly to the study that was presented in the previous chapter, user perception of system dynamics that might occur during a video call in heterogeneous wireless networks are addressed in this chapter. The presented work is based on several user tests that were designed to analyse the Quality of Experience in wireless networks, such as WiFi and HSDPA, where either robust or high fidelity video codecs are available and whose encoding bit rate can be adjusted. In particular, it is analysed how the effect of switching between those networks, codecs, and bit rates is perceived during an ongoing video call and whether those techniques can be used to adapt the transmission quality. Similarly to the study that was presented in the previous chapter, the results are used to draw guidelines for a perception-based mobility management, and to analyse how well the existing quality predictions models can estimate the perceived quality in the context of time-varying transmission that is expected in heterogeneous wireless networks.

5.1 Experimental Design

The design of the user experiments that were carried out in this study was guided by a similar methodology to that already presented in the previous chapter. It is based on viewing of short signal samples and on emulation of video calls. However, a conversational test was not performed in this case, as it was shown in the previous chapter that such an interactive test introduces certain research disadvantages. First of all, during a real conversation users are generally less analytic in their judgments and subtle perceptual differences get blurred. Moreover, instable quality during a conversation turned out to be challenging to harvest reliable results. The more dynamics and parameters are investigated in a test, the greater is the risk of data unreliability. In case of a video call, however, the number of the analysed parameters is greater than in case of speech telephony. Simultaneous transmission of audio an video increases the number of test combinations. Therefore, a conversational test was not conducted

B. Lewcio, *Management of Speech and Video Telephony Quality in Heterogeneous Wireless Networks*, T-Labs Series in Telecommunication Services,
DOI: 10.1007/978-3-319-02102-7_5, © Springer International Publishing Switzerland 2014

during this study, and the work presented in this chapter is based on the remaining methodologies that enable to get more analytic insights in the user perception. Emulation of video calls and viewing-only experiments were performed to analyse the perception of video quality in heterogeneous wireless networks and at the same time to capture more granular results. As a result, the conducted tests were subdivided as follows:

- One test addressed the quality of short video-only samples. Test V1 was used to get a general insight in the perception of video quality in heterogeneous wireless networks.
- One test was focused on the quality of short audiovisual samples that are typical for video telephony. Test V2 was focused on the perception of major effects that are relevant for this type of service in heterogeneous systems.
- One test was designed to analyse the quality of emulated video calls. Test V3 was used to explore user perception of quality in a laboratory situation that is close to a real video call in heterogeneous wireless networks.

The above selection of user tests enabled to understand how a single effect, i.e. a single change of quality, is perceived, and what is its effect on the overall experience of a video call. Moreover, the subdivision of the viewing experiments into a video-only and an audiovisual test enabled to analyze the quality perception in two different modalities.

Test Procedures

Test V1 and V2 were carried out according to ITU-T Rec. P.910 and P.911 [1, 2] that describe the standardized procedures of video-only and audiovisual viewing tests, respectively. In both of the experiments, approx. 9s long test sequences were presented to the invited participants who judged the video quality in Test V1, and audio, video, and audiovisual quality in Test V2 on the recommended continuous 11-point MOS scale. The ratings were collected during two sessions of approx. 30 or 40 min in Test V1 and Test V2, respectively. Both of the sessions in each test were separated by a 10 min long break. Prior to each of the experiments, the test participants followed a training phase during which they judged a set of 15 stimuli that approximately covered the range of addressed test conditions.

The third of the conducted tests, Test V3, was designed in analogy to the call quality test that was presented in the previous chapter. It was based on a test procedure that is recommended in ETSI TR 102 506 v.1.2.1 [3] that is desired for assessment of speech telephony quality, and that was validated for the video telephony use case in this work [4]. The main principle of this methodology describes how to create a test situation in which a user focuses more on the content of a presented test stimulus than on its quality. To fulfill this requirement, Test V3 consisted of multiple emulations of video calls that were presented to the invited test participants in two sessions of approx. 45 min with a 10 min break in between. Each of the calls was constructed of five, approx. 9s long, samples that were concatenated in a meaningful order to tell a story that is typical for a video call. After each of the first four samples were presented, the test participants were asked a content-related question, having the same amount of time to orally provide an answer. After the last sample of an emulated call was played-out, the participants judged the overall quality of the video call on the continuous 11-point MOS scale. As a result, a total duration of 80s per call emulation was observed. In order to familiarize the test participants with the test procedure and the quality range that was included in the test, a training phase was introduced prior to the experiment.

All of the described experiments were conducted in a test room that was conform to ITU-T Rec. P.911. The test material was displayed in a randomized order on a Dell Inspiron 1012 notebook with a 10.1 inch WSVGA (1024 × 600 pixels) display. The choice of a notebook as a test device was determined by the screen size that was close to that of a mobile device and by a hardware architecture that is supported by Linux and by Microsoft Windows. The support of both operating systems was required by the proprietary software that was used in this work. The notebooks were used by the test participants who were sitting in front of the screen with a viewing distance of approx. 50 cm. In case audiovisual content was provided, the audio signal was presented using a Sennheiser HMD 410 headset at a comfortable listening level with a background noise below 35 dB(A). 21 participants took part in Test V1, and 20 test users attended Test V2 and V3. The test participants were recruited from the university campus and were balanced in gender. They were non-experts as they were not concerned with multimedia quality as part of their work. They were interviewed about hearing and visual disorders and the effort of completing a tests was rewarded with a shopping voucher.

Source Material

To generate the test samples that were used during the tests, two sets of source material were used. A common set of source samples was shared in Test V2 and V3, but a different pool of video clips was applied in Test V1. The source material of the video-only Test V1 included three different content types that were represented by four different video sequences. Video 1 presented a 10s long extract of a movie, Video 2 included a 10s long extract of a soccer game, and Video 3 and 4 presented a men and a women participating in a video call. The last two "head and shoulder" clips were extracted from the reference set of Test V2 that is described in the next paragraph. In general, the choice of the videos in Test V1 was made according to the typical content that is delivered in mobile networks, and to empirically prove that the content itself has a significant impact on the Quality of Experience in heterogeneous networks [5].

Accordingly, the source material of Test V2 and V3 was based on a pool of video clips that presented two male and two female participants of a video call. The speakers were recorded while accomplishing a certain task. For example, it was the goal of a video call to order a washing machine, to rent a car, or to make an appointment with a doctor, which was consistent with the study that was presented in the previous chapter. While performing a task the speaker was taking an advantage of the video communication channel by showing content-related items, such as pictures or leaflets, to the camera. Each of the story-tellings was split into 10 samples of approx. 9s, which generated a pool of 40 audiovisual samples that are typical for telephony. Subsets of only five interrelated samples were used to emulate each of the video calls in Test V3.

The entire pool of the reference material from all of the tests was stored in the VGA format with 25 frames per second (fps). The resolution of 640 × 480 pixels was considered as the upper bound of real-time communication that was supported by mobile computing at the time the project was started. This trend has changed over the years, but the VGA resolution is still a popular format that is used in mobile video telephony systems. The audio format of the reference material was stored as uncompressed wideband, which was consistent with the work that was presented in the previous chapter.

Test Conditions

The source material was used to produce the test sequences that were used during the conducted experiments. To generate the test data, a set of core transmission parameters was defined in this study (cf. Table 5.1). Two wireless and one wired network, HSDPA, WiFi, and LAN were addressed. The selection of the wireless networks was motivated by a trade-off between the coverage and the link-layer quality that is represented by both technologies. While the coverage area of a WiFi network is relatively small as compared to that of HSDPA, WiFi provides a faster and a more reliable wireless connection. LAN was applied as a reference network, and it was used for an additional qualitative benchmark of the wireless technologies in the initial stage of the project only.

The networking circumstances were deliberately impaired in selected cases, and to do this random packet loss up to 5 % was emulated. The maximal degree of degradation was determined according to some preceding tests that showed that the quality of a VGA clip with the frame rate of 25 fps was already rated at the bottom of the scale. The degradation was an effect of visible decoding errors rather than of video freezing. However, the latter occurred due to jitter buffer underruns too.

In all of the test networks two video codecs, H.264 and MPEG-4 [6], were used. They were selected because they are commonly used in mobile communication systems, and their open-source implementation [7] introduced a trade-off between the compression quality and packet loss robustness. A video stream that was encoded with MPEG-4 was more robust against packet loss, although H.264 provided higher quality video. Moreover, also the compression artifacts were different between the

Table 5.1 Transmission parameters addressed in the conducted audiovisual tests

Parameter	Value
Network	WiFi, HSDPA, LAN
Packet loss	0, 1, 3, 5 %
Video codec	H.264, MPEG-4
Video bit rate	256, 768, 1536 kbit/s
Frame rate	25 fps
Speech codec	AMR wb. 23.05 kb/s, AMR nb. 12.2 kb/s

codecs, and the effect was in particular noticeable when the encoding bit rate was low. In such a case, MPEG-4 caused visible blockiness of the video, while the H.264 signal was blurred; it is recognised that H.264 is a more efficient video codec that provides more space for sophisticated means of error control. However, this study addresses solely the trade-off between efficient and robust signal compression. Benchmarking of diverse techniques of error control is out of scope of this work.

Both of the codecs were configured to compress the native format of the source material and no rescaling was applied. The encoding bit rate was configured to 256, 768, or 1536 kbit/s, which was selected to cover the low, medium, and high quality of video, and to emulate various loads of a wireless connection. Moreover, the codecs were used as provided by the library without any additional tuning of the encoding parameters, nor an additional packet loss concealment was applied. This configuration might not ideally reproduce real telecommunication systems, where a variety of proprietary solutions are used to improve the video quality that is provided by commercial products, but the access to such proprietary solutions is limited for the research community. The use of an open-source codec library, in turn, enabled to produce results that are publicly comparable.

Finally, when audiovisual quality was addressed, the speech signal was encoded with either ITU-T G.722.2 aka "AMR wideband" or 3GPP TS 26.073 aka "AMR narrowband" that are two speech codecs which are commonly used in mobile networks [8, 9]. The AMR wideband codec provides wideband speech (50–7000 Hz) at encoding bit rates between 6.6 and 23.85 kbit/s, and AMR narrowband offers narrowband speech quality (300–3400 Hz) at the bit rates between 4.75 and 12.2 kbit/s. However, only the bit rate of 23.05 and of 12.2 kbit/s for wideband and narrowband transmission, respectively, were selected in this work, which assured the best encoding quality amongst all of the available encoding modes that are supported by both of the codecs.

The above described transmission parameters were used to design the test cases of the conducted experiments. It has to be emphasised that the general restriction that was introduced in the previous chapter that if a network is switched it will be required to adapt the codec to new networking circumstances was released. As a result, a more systematic design of the test conditions was possible and the isolated impact of individual effects, such as a network, codec, or bit rate switch, on user perception could be addressed. Nevertheless, due to the number of the addressed parameters and switching combinations resulting thereof, not all of the conditions of interest could be included in a single test. Therefore, a careful selection of the test cases according to the focus of an experiment was required, and the duration of a test was extended to perform a systematical analysis.

As a result, the test material of the first two passive tests, Test V1 and Test V2, consisted of 50 test conditions that formed a balanced test plan to cover the entire range of the quality scale. All conditions were grouped in stable and dynamic conditions. The stable conditions represented various configurations of constant video streaming and the dynamic conditions included a change of a transmission parameter in the middle of a sample. Network handover, video codec changeover, video bit rate switching, and speech codec changeover between wideband and narrowband quality were considered as relevant for mobile service delivery. Both condition groups were processed in a clean and a degraded transmission channel so that transmission impairments due to packet loss and perception of quality adaptation was addressed.

The purpose of both of the short sample tests was different. Test V1 was a video-only test that was carried out to explore the main effects that are perceived during a wireless transmission in heterogeneous networks. Therefore, all of the aforementioned parameters, excluding the speech codecs, were used to design the following five groups of test conditions (cf. Table A.17):

- 15 reference conditions that represented constant video transmission and that were meant to be anchor conditions of the test (Conditions 20–34).
- 10 conditions that targeted user perception of video quality during a network handover between WiFi and HSDPA, which was investigated for various video codec and bit rate configurations (Conditions 1–10).
- 6 conditions that addressed perception of video quality during a codec changeover between H.264 and MPEG-4, which was analysed for various encoding bit rates in WiFi that was considered as the reference wireless network of this work (Conditions 35–40).
- 10 conditions that were designed to address user perception of video quality when the encoding bit rate is switched. This effect was explored for all combinations of the video bit rates and codecs in WiFi (Cond. 41–50).
- 9 conditions that were focused on user perception of quality adaptation. In this case, a video stream that was encoded with H.264 at 1536 kbit/s in WiFi and that was affected by packet loss, was adapted by a change of the codec to MPEG-4, by reduction of the encoding bit rate to 256 kbit/s, or by a handover to a loss-free HSDPA network (Conditions 11–19).

Each of the conditions was generated for all four source sequences that were described before in this section.

As a result, Test V1 broadly covered the spectrum of the relevant transmission parameters in heterogeneous networks and it laid a foundation for the next tests. Based on the first results, the set of transmission parameters was refined. The medium video bit rate of 768 kbit/s and the LAN network were removed from the next analysis, because they were not causing any unique effects. Due to the same reason a network handover in the direction from HSDPA to WiFi was no longer addressed. As a replacement of the removed conditions, new test cases that address the influence of the speech quality on the overall experience of the audiovisual quality were included in Test V2.

The refined design of Test V2 resulted in six groups of test conditions (cf. Table A.19):

- 19 reference conditions, whereof 15 conditions were produced for wideband and four conditions for narrowband speech quality. User perception when high speech quality is provided was addressed most extensively in this work (Conditions 1–15, and 34–37).
- 4 network handover conditions that were generated for wideband speech quality only (Conditions 16–19).

- 4 video codec changeover conditions that were generated for wideband speech quality only (Conditions 20–23).
- 4 video bit rate switching conditions that were generated for wideband speech quality only (Conditions 24–27).
- 4 conditions that addressed used experience during a changeover between wideband and narrowband speech codec, when high or low video quality was provided (Conditions 38–41).
- 15 conditions that were used to analyse user perception of quality adaptation when the connection quality is degraded. In this case, nine and six conditions were produced for wideband and narrowband speech, respectively (Conditions 28–33, and 42–50).

Each of the test conditions was generated for four source samples, which was one random sample that was recorded for each of the speakers, assuring that all speakers are equally represented in the test set.

The conditions from Test V2 were used to create call quality profiles that were analysed in Test V3, which created a link between the perception of selected transmission extracts and the overall quality of a call. Moreover, the quality profiles that were used in Test V3 were designed to fulfill two further research purposes. On the one hand, they were aimed to emulate degradation patterns that are expected during a video call in heterogeneous networks. On the other hand, the profiles were created to allow further development of call quality models for video telephony. Therefore, apart from profiles that emulate mobility, also profiles that introduce quality degradation at the beginning, in the middle, and at the end of a call, as well as profiles with constant, raising, or decreasing quality, were designed. As a result, the following three groups of profiles were introduced (cf. Table A.21 and Figure A.4):

- 5 profiles that addressed user perception of constant quality during a video call. Low, medium, and high quality profiles were realized. Those profiles were meant to be anchor conditions of this test (Profiles 1–4, and 11).
- 6 profiles that targeted user experience of mobility. Leaving or joining an access point was emulated by changing the packet loss rate, and as the case may be, a network, codec, or bit rate switch that was performed to adapt the quality in the middle of such a call. Also the isolated effect of a network handover in the middle of a call was addressed (Profiles 4–10).
- 5 profiles that were focused on the impact of the temporal position of a quality change on the overall experience of the video call quality. For this purpose, the video bit rate was increased at the beginning (first sample), in the middle (third sample), or at the end (fifth sample) of an emulated call, the bit rate was decreased at the end, or the middle sample of a call was substantially degraded (Profiles 12–16).

Each of the profiles was generated for all four speakers from the aforementioned pool of source sequences.

The test material that was described in this section was generated in the heterogeneous testbed that was created in this project (cf. Chap. 3). The uncompressed source material was fed to the Correspondent Node, and then the signal was in real time encoded, transmitted, de-jittered, decoded, and recorded on the Mobile Node. This procedure enabled to capture the video signal that would be otherwise directly displayed to the user. Moreover, some other settings of the testbed that were used during this study have to be mentioned. To preserve low-delay transmission that is typical in a video telephony system, the jitter buffer of the video client was configured to the boundaries of 80–200 ms, and the structure of the encoded Group of Pictures was configured to IPPP [10]. The B frames were not used during encoding to reduce the system delay. Moreover, both, codec and bit rate switching, were realized with the same technique that is described in Sect. 3.3. An implementation of scalable video coding (SVC) [11] was not integrated in the testbed, as it was not supported by the open-source library that was used in the project. Finally, the size of the transmitted RTP packets was limited to 1400 Bytes to avoid any payload fragmentation in the Mobile IP tunnel.

5.2 Quality Perception

The data that was collected during the subjective experiments was analysed in analogy to the evaluation that was presented in the previous chapter. First of all, the user judgements were filtered in terms of their reliability (cf. Table A.15), and the results that were generated by unreliable test participants were removed from the data base. The filtered data was then averaged per condition and the results were linearly mapped on the 5-point MOS scale that is a common scale that is used in this work; the detailed results of all the tests can be found in Appendix A.2.

The analysis in this section is organized according to the test structure that was presented in the previous section. The quality of short video-only stimuli is analysed first, and quality of the audiovisual samples is analysed next. The study of the quality perception during emulated video calls is presented as last in this section. Moreover, the individual tests are analysed according to the condition groups that were described above. For example, user perception of quality during a network, codec, or bit rate switch is evaluated separately. Cross-test observations are provided chronologically at the relevant positions in the text as the evaluation evolves, as well as in the summary of the entire section.

5.2.1 Quality of Short Video-Only Samples

The impact of the network technology on the perceived quality in Test V1 was primarily analysed when the encoding bit rate was high, so when the major perceptual effects were expected. The remaining bit rates were addressed in an exploratory way. As a result (cf. Fig. 5.1), if the entire video transmission was performed in one network only, the perceived quality was always higher in WiFi than in HSDPA. This difference

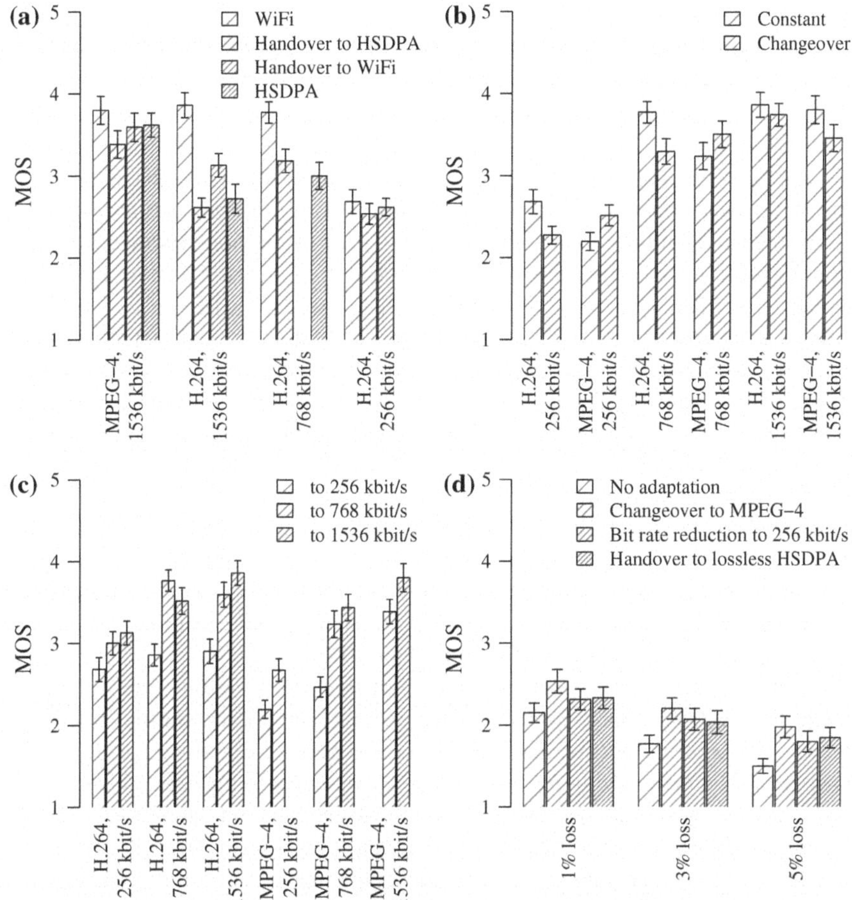

Fig. 5.1 User judgements of quality in Test V1. MOS_v, and 95 % CI. The x-axis describes the initial state and the switch of the addressed parameter is performed while preserving the remaining parameters constant. If both states are equal in case of bit rate switching, the transmission remains constant. Selected groups to address user perception of: **a** Network, **b** Video codec, **c** Video bit rate, **d** Transmission adaptation due to packet loss

between the networks was substantial when the H.264 codec at the medium or high bit rate of the test was used only. When MPEG-4 was used, smaller differences between HSDPA and WiFi were observed than in the former case. Apparently, H.264 was more sensitive to the underlying transmission technology than MPEG-4. Additionally, as a side track of the analysis, the quality provided in WiFi was compared with that measured in the wired LAN network (not represented in the figure, cf. Table A.18). The video quality in both of the networks was perceived similarly, which means that the Quality of Experience in the reference wireless network of this study was comparable with that in wired systems.

When a network handover in the middle of a test sample occurred, the Quality of Experience was always affected. The perceived effect was related to the handover direction, the utilized video codec, and to the encoding bit rate. The higher was the video bit rate and the less robust the video codec, the stronger was the perceived effect. At the highest bit rate of H.264, a network handover from WiFi to HSDPA degraded the video quality to a level that was considerably lower than that in HSDPA. A handover in the opposite direction, from HSDPA to WiFi, provided a substantial improvement of quality as compared to pure HSDPA transmission. Also at the highest bit rate of MPEG-4, when the network was switched from WiFi to HSDPA, the quality was rated considerably worse than that in HSDPA. However, when the network was switched from HSDPA to WiFi an improvement of the perceived quality was not observed and the user experience remained similar to that in HSDPA. As a result, it was once again confirmed that H.264 was more sensitive to networking circumstances than MPEG-4. When the medium encoding bit rate of the test was applied, not all of the relationships could be explored due to the lack of relevant test conditions. However, it was confirmed that a handover to HSDPA considerably degraded the quality when both, H.264 and MPEG-4, were in use. Finally, at the lowest encoding bit rate of this test the impact of a handover was never substantial. The strongest degradation in this case was once again observed when it was switched from WiFi to HSDPA and the H.264 codec was used. As a result, it was confirmed that H.264 was less robust to connection degradations than MPEG-4.

In sum, the above presented results showed that when various types of video content (cf. Sect. 5.2.2 for a content-related comparison) are transmitted in heterogeneous networks, wireless technology and a network handover affect the Quality of Experience. The perceived impact of a handover is related to the network switching direction, robustness of the video codec, and to the encoding bit rate. If the encoding bit rate is high, the use of a stable network connection is advantageous, and a network handover is in general not recommended, although a handover to a stable WiFi network might help to improve user experience when a high-fidelity video codec is used in an unstable environment. Moreover, the negative effect of a handover or of an instable network transmission is reduced when a robust codec is used or when the video bit rate is low.

As it was empirically proven above, MPEG-4 was more robust against network degradations than H.264. However, if stable transmission quality was assured, H.264 always provided higher video quality than the other codec. This difference between the codecs was related to the encoding bit rate. While similar quality ratings were observed for both codecs when the encoding bit rate was high, the quality of video that was provided by H.264 was substantially higher than that encoded by MPEG-4 in any other case. This result confirms the study from [6] that explains how the design of H.264 was optimized for low transmission bit rates.

When the video codec was switched in the middle of a test sample, two trends were observed. When the video bit rate was high, changeovers between H.264 and MPEG-4 degraded the perceived quality below the quality level that is provided by pure MPEG-4 transmission. Unexpectedly, a switch from MPEG-4 to H.264 was rated much worse than a changeover in the opposite direction. This effect, however, was related to the design of the codec changeover solution, and the details of this problem are explained in Sect. 3.3.3. When the encoding bit rate was medium or low, the experience of a changeover was rated between the qualities that were provided by both of the codecs. A changeover from MPEG-4 to H.264 always improved, and a changeover in the opposite direction always degraded the quality perception. The quality improvement was less perceived than a quality degradation that, in turn, always resulted in a quality rating that was close to that collected for the lower quality codec. Further analysis of the test data disclosed that the effect of a changeover was stronger perceived when the video bit rate was low, which was related to the aforementioned fact that the H.264 codec was optimized for low transmission bit rate, and hence a notable quality change during a switch was observed. In this case, switching the codec in both directions improved or degraded the user experience substantially.

In sum, robustness and efficiency of a video codec must be matched for networking circumstances to maximize the Quality of Experience. This decision is of particular relevance when the transmission bit rate is limited or the networking conditions are impaired. Moreover, switching between video codecs during an ongoing transmission in general improves or degrades the quality according to logical expectations. Doing so, a switch to a low quality codec has a remarkably stronger impact on user perception than a changeover in the opposite direction. Therefore, frequent codec changeovers should be avoided, and if a quality degradation is expected, constant use of the low quality codec enables to reduce the number of quality management operations at no substantial loss of perceived quality. An exceptional case of the presented trend was discovered at the highest encoding bit rate of the test. Due to the implementation of the codec changeover solution, any codec changeover in this case degraded the perceived quality.

The above observations disclosed that the video bit rate was one of the key parameters that influenced the video quality in Test V1. Further analysis of this finding showed that the influence was substantial. However, when the video bit rate was increased, the quality improvement was not proportional. This means that the higher the encoding bit rate was, the lower was the relative improvement of the perceived video quality.

When the bit rate was switched during an ongoing session, the end quality of a sample was changed along with the bit rate switching direction. An exception of this trend was observed when H.264 was used and the encoding bit rate was switched from 768 to 1536 kbit/s. In this case a quality degradation was observed, which, however, was once again related to the design of the codec changeover solution that was used in this study to switch the encoding bit rate. If the bit rate was switched from the lowest to any of the higher bit rates, the Quality of Experience was always substantially improved, and the improvement was related to the end bit rate. The effect was also more noticeable when MPEG-4 was used, which once again confirms that MPEG-4 is less efficient at low transmission bit rates than H.264. When the bit rate was switched from any of the higher to the lowest one, a substantial quality drop was observed. The quality in this case was judged close to that provided by constant transmission at the lowest bit rate. This finding is in line with previous results, which confirms that a quality degradation is perceived stronger than a quality improvement. A similar effect was observed when the bit rate was switched between the highest and the medium bit rate of the test, excluding the aforementioned exception that was observed when H.264 was used.

In sum, high encoding bit rate helps to maximize user experience of video quality, but it has to be emphasised that the relationship between utilized resources and the perceived quality is not proportional. This means that the higher the encoding bit rate is, the lower is the relative improvement of the perceived video quality. Moreover, switching the bit rate during an ongoing transmission is in general perceived according to logical expectations. However, it has to be taken into account that degradation of quality is perceived stronger than a quality improvement. Therefore, if a drastic reduction of quality is expected, proactive utilization of a constantly low bit rate might reduce the number of quality management operations, while preserving the Quality of Experience.

As an additional track of the presented analysis, all of the above techniques were compared in terms of their effectiveness to improve the perceived video quality when transmission degradations occur. As a baseline scenario, a H.264 stream at the bit rate of 1536 kbit/s in WiFi that was affected by packet loss values of 1, 3, or 5 % was used. The transmission was adapted by a handover to a loss-free HSDPA network, by a changeover to a more robust MPEG-4 codec, or by a reduction of the encoding bit rate to 256 kbit/s.

When the transmission was not adapted and the original video stream was permanently affected by packet loss, the user experience of quality was substantially degraded. The impairment exhibited an exponential trend, which means that the relative degree of degradation was reduced when the packet loss was increased. However, the Quality of Experience was always improved when one of the adaptation techniques was applied. When the codec was changed to MPEG-4, the improvement was always substantial. When the bit rate or network were switched, the improvement was substantial only in case the quality degradation was higher than 1 % of packet loss. A post-hoc analysis (Scheffé) has shown the quality judgements that were collected for the network handover and the bit rate reduction were similar to each other ($p = 0.97$), and that the codec changeover significantly outperformed the other techniques of quality adaptation (resp. $p = 0.007$, $p = 0.004$).

In sum, it is of benefit to use all of the analysed techniques of transmission adaptation when packet loss occurs. However, a changeover to a more robust codec turned out to be a technique that always improved the quality substantially. The other methods generated a substantial quality gain only when the connection degradation was higher than 1 % of packet loss. This indicates that a bit rate reduction is not an efficient way of quality adaptation in wireless systems. Finally, it has to be emphasised that if a mobile user is leaving a WiFi area, a network handover is the only possibility to maintain the connection.

The above presented analysis enabled to explore the major effects of video quality perception in heterogeneous wireless networks, which helped to draw some initial guidelines for mobility management in those networks. It was shown that proper selection of the transmission network, video codec, and the transmission bit rate is essential to provide high Quality of Experience in heterogeneous wireless networks. As a matter of fact, if high quality of video should be provided, stable network connection, high quality codec, and high encoding bit rate are required. However, it has to be considered that although the encoding bit rate enables to maximize the Quality of Experience, the perceived quality improvement is not proportional. The higher is the encoding bit rate, the lower is the relative quality improvement. Moreover, any transmission changes during a high quality streaming are not recommended, although it might be advantageous to switch the network when a connection is initially degraded. The situation is different when a transmission environment is constrained. In such a case, proper match of the video codec for networking conditions is of particular relevance. When the connection throughput is limited, an efficient coding scheme can help to substantially improve the Quality of Experience. In turn, a robust video codec is required if an unimpaired network connection cannot be guaranteed. Coding robustness and low encoding bit rate are the two parameters that help to reduce visible video artifacts, including that which might occur due to a network handover.

When the quality is adapted during an ongoing session, switching between codecs or bit rates improves or degrades the quality according to the logical expectations. However, it has to be considered that a reduction of the video quality has a stronger impact on the user perception than a quality improvement. Therefore, if a drastic reduction of quality can be foreseen, proactive utilization of a low quality configuration reduces the number of service management operations at no substantial costs of the perceived quality. Moreover, quality adaptation is always advantageous when a transmission is affected by packet loss. In this case, the most noticeable improvement can be achieved by replacing the video codec with a more robust encoding scheme, but it has to be emphasised that if a mobile user is leaving a WiFi area, a network handover is the only possibility to maintain a connection. Therefore, if a connection is degraded it is recommended to adapt the quality by changing the video codec first, and to switch the network when a certain level of degradation is exceeded.

5.2.2 Quality of Short Audiovisual Samples

The analysis that was presented in the previous subsection was not strictly designed for video telephony systems. The influence of the audio signal on the overall experience of quality was not addressed, and the evaluation of the video quality perception was based on various content types that are typically delivered in mobile networks. The content, however, influenced the video quality significantly ($F = 71.2$, $p < 0.001$). Therefore, to gain a more detailed insight in the perception of time-varying quality during a video call, Test V2 was carried out.

In opposite to the results of the previous test, Test V2 showed (cf. Fig. 5.2) that when the "head and shoulders" content was transmitted in one network only, the Quality of Experience was not substantially affected by the wireless technology. Even though a considerable impairment of quality was denoted in HSDPA when

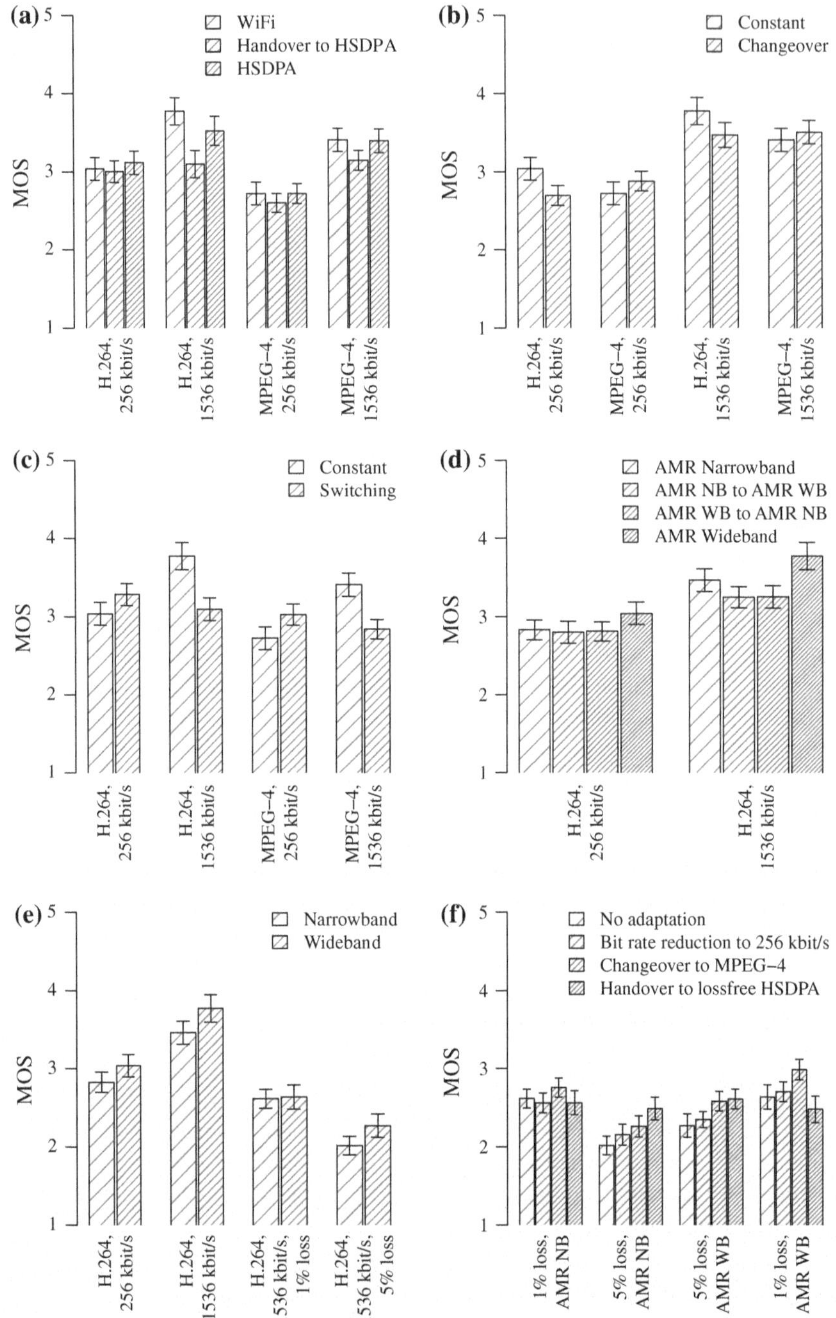

Fig. 5.2 User judgements of quality in Test V2. MOS_{av}, and 95 % CI. The x-axis describes the initial state and the switch of the addressed parameter is performed while preserving the remaining parameters constant. Selected groups to address user perception of: **a** Network, **b** Video codec, **c** Bit rate, **d** Speech codec, **e** Audiovisual trade-off, **f** Transmission adaptation due to packet loss

H.264 at high encoding bit rate was used, this degradation was lower than in Test V1. When the network was switched from WiFi to HSDPA in the middle of a sample, the video quality was always degraded and the overall user experience was always worse than that in both of the networks. The quality degradation was perceived stronger when the video bit rate was high, but a substantial effect was only observed when H.264 was used. Therefore, it was once again confirmed that H.264 was more sensitive to networking impairments than MPEG-4.

As a result, when stable transmission in either HSDPA or WiFi can be guaranteed during a video call, switching between these networks is not recommended. If the necessity of a handover can be predicted, proactive scheduling of the entire transmission in a lower quality network allows to maximize the user experience and the number of service management operations is simultaneously reduced. In any other case, robust video compression and low encoding bit rate help to mask visible artifacts that are perceived due to a handover.

Similarly to the analysis in Test V1, the Quality of Experience that was provided by both of the addressed video codecs, H.264 and MPEG-4, in a stable WiFi network was compared next. The results confirmed that H.264 always provided substantially higher Quality of Experience than MPEG-4. Accordingly, when the video codec was switched during an ongoing session, the end quality of such a sample was always judged between the ratings that were collected for MPEG-4 and H.264, and the quality perception was changed according to the logical expectations. A changeover from MPEG-4 to H.264 always improved, and a switch in the opposite direction always degraded the Quality of Experience. Doing so, the quality improvement was less perceived than the quality degradation. In the latter case, the overall quality was rated close to that provided by the lower quality codec.

In comparison to the results of Test V1, two noticeable differences were observed in Test V2 when the video codec was switched from MPEG-4 to H.264. First, when the encoding bit rate was low, the resulting quality improvement was considerably low in Test V2. Second, when the encoding bit rate was high, a changeover in Test V2 did not degrade the perceived quality; no explanation of these differences could be found.

As a result, the video codec must be selected according to the networking circumstances of a video call. If a connection is impaired, the use of MPEG-4 is recommended, but if a stable network is available or the connection throughput is limited, H.264 should be used to maximize the Quality of Experience. Moreover, switching between codecs during a video call is perceived according to the logical expectations, and hence a switch to H.264 always improved the user experience. However, it should be considered that a degradation of quality has a remarkably stronger impact on the user perception and if a switch to a low quality codec can be foreseen, constant use of the low quality codec enables to reduce the number of quality management operations at no substantial loss of perceived quality. As a matter of fact, those results confirm the main trends that were discovered in the previous test.

Accordingly, the video encoding bit rate was also one of the main parameters that influenced the Quality of Experience in Test V2. The higher the encoding bit rate was, the better was the overall quality. When the bit rate was switched during an ongoing transmission, the user experience was changed according to the logical expectations. When the video bit rate was increased, the overall quality impression was improved. However, the improvement was only substantial when MPEG-4 was used, which is once again related to the fact that this codec is less optimized for low transmission bit rates than H.264. When the encoding bit rate was reduced, the quality perception was always substantially degraded, and the overall quality of such a sample was close to that observed for constant use of the lower bit rate. Therefore, the effect of a bit rate reduction was stronger perceived than that when the bit rate was increased.

As a result, the encoding bit rate should be adjusted according to the connection throughput to maximize the Quality of Experience. This is in particular important for codecs such as MPEG-4 that are not efficient at low transmission bit rates. Moreover, adaptation of the encoding bit rate in a call is perceived according to the logical expectations, but once again it should be noted that degradation of quality has a remarkably stronger impact on user perception than a quality improvement, and hence oscillation of the delivered quality should be avoided.

Test V2 was also used to analyse the impact of speech quality on the overall Quality of Experience. In particular, the influence of the speech encoding bandwidth on the overall quality of an audio-visual sample was addressed. The results confirmed that high speech quality improved the user experience. In fact, the speech quality was more relevant, when high quality video was provided. When the bandwidth of the speech signal was changed in the middle of a sample, the user experience was always degraded, and the quality of those samples was rated worse than that of pure narrowband. This effect was independent from the switching direction, but it was stronger perceived when the video quality was high. This observations are consistent with the results that were presented in the previous chapter, which confirms that the previously discovered trends also hold in context of video telephony.

Additionally, the quality scores that were collected in Test V2 were evaluated in terms of the cross-correlation between the audiovisual, audio, and video quality. The results showed (cf. Table 5.2 and Fig. 5.3) that the collected audiovisual quality scores were more correlated with the video than with the speech quality, although the speech quality also substantially contributed to the overall Quality of Experience.

As a result, high quality of the speech signal improves the Quality of Experience during a video call, and the speech quality matters more when the video quality is high. Moreover, switching between wideband and narrowband speech encoding during a video call should be avoided as much as possible, and alternatively constant narrowband speech quality should be provided. However, if the transmission must be adapted, it is recommended to change the video configuration first, before the

Table 5.2 Correlation of the audio, video, and audiovisual judgements in Test V2

	MOS_{av}	$MOS_{a\|av}$	$MOS_{v\|av}$
MOS_{av}	-	0.687	0.931
$MOS_{a\|av}$	-	-	0.396
$MOS_{v\|av}$	-	-	-

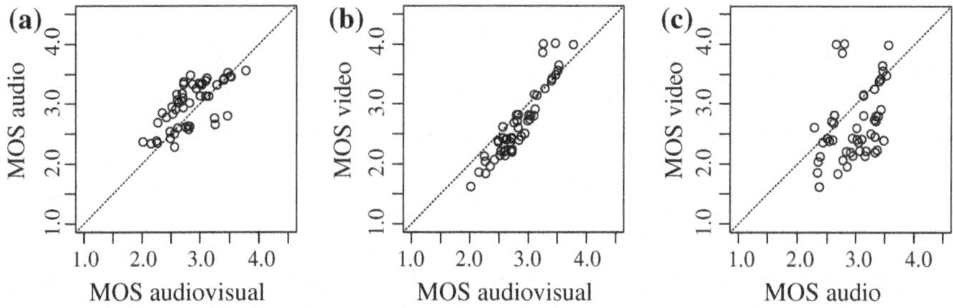

Fig. 5.3 Comparison of user judgements of individual quality modalities in Test V2. **a** Audio and audiovisual, **b** Video and audiovisual, **c** Video and audio

bandwidth of the speech signal is switched. Low video quality reduces sensitivity to degradations that appear in the audio channel.

Finally, all of the above techniques of video quality adaptation were analysed in terms of their efficiency when the quality of a "head and shoulders" sample was degraded. As a baseline scenario an audiovisual stream that was encoded with H.264 at 1536 kbit/s and with either AMR wideband or with AMR narrowband codec and that was transmitted in a WiFi network that was affected by either 1 or 5 % of packet loss was used. The transmission was adapted by a handover to a loss-free HSDPA network, a changeover to a more robust MPEG-4 codec, and by a reduction of the encoding video bit rate to 256 kbit/s in the middle of a sample.

The results confirmed that a changeover to MPEG-4 was the only technique that always improved the Quality of Experience. The stronger the packet loss was, the more observed was the quality improvement. The achieved effect of a changeover was additionally amplified when the quality of the speech signal was high as well. When the video bit rate was reduced to counterbalance the effect of packet loss, in most of the cases the QoE was slightly improved, although a quality degradation was observed when the packet loss was low and narrowband speech quality was provided. This observation confirms that a bit rate reduction is not an efficient technique of quality adaptation in wireless networks, when transmission-related degradations occur. Finally, if the network was switched to HSDPA to combat the packet loss, the Quality of Experience was degraded if the loss rate was low, but a quality improvement was observed if packet loss was high. Moreover, the perceived degradation due to a handover was amplified if the speech quality was high. Therefore, according to the results of this test a hypothesis can be derived that users are more sensitive to adaptation of video quality when the speech quality is high.

An additional comparison of the above results with quality ratings that were obtained when the signal was transmitted in a loss-free HSDPA network or when MPEG-4 was permanently used in a degraded network (cf. Table A.19) revealed that these conditions were rated higher than the corresponding methods of quality adaptation above. Therefore, if quality degradations are expected, proactive selection of one of the robust transmission configurations improves the Quality of Experience.

As a result, it was confirmed that also in the context of video telephony a changeover to a more robust video codec turned out to be the adaptation technique that always improves the Quality of Experience. However, if packet loss is high, network handover is the most recommended technique of transmission adaptation, which is of particular relevance when a mobile user is leaving a WiFi area. Therefore, it is once again recommended to change the video codec first, and to switch the network when a certain level of degradation is exceeded. As a matter of fact, it was also shown that the control of the video encoding bit rate is not an effective technique of transmission adaptation when link-layer impairments occur. However, if the instability of a connection can be foreseen, constant use of either a robust video codec, such as MPEG-4, or permanent transmission in a loss-free but lower-quality network, such as HSDPA, helps to proactively deliver higher Quality of Experience than any other of the analysed quality adaptation methods.

As an additional side track of the analysis in this section, the judgements of the video quality that were collected in Test V1 and V2 (MOS_v in Test V1 and MOS $MOS_{v|av}$ in Test V2) were compared with each other. Equal video conditions from both tests were matched, which eventually included test conditions that were created for two different audio configurations in Test V2. As a result, a list of 41 comparable conditions was created. The results of the analysis revealed that the video judgements of both tests were in 94.6 % correlated with each other, and the difference expressed as an RMSE value was 0.26 MOS. An additional ANOVA analysis that was based on a linear model with the difference of the MOS scores between the tests as a dependent variable and the test factors of Test V2 as independent variables revealed that the speech codec ($F = 0.54$; $p = 0.659$) did not contribute to the difference between the video quality ratings in both tests. The encoding bit rate was a slightly significant factor ($F = 2.71$; $p = 0.066$), and the video codec ($F = 9.154$; $p < 0.001$), as well as wireless technology ($F = 4.126$; $p = 0.29$) were the main reasons of the observed difference.

The analysis that was presented in this section enabled to refine the quality management recommendations that were derived in the previous test. Most of the trends that were discovered in Test V1 were confirmed in Test V2, although some differences between the results of both tests were revealed. Apparently, less dynamics in the source material that was used in the video telephony related test improved the QoE in heterogeneous networks. Accordingly, when a video call is performed in either WiFi or HSDPA only, the wireless technology does not influence the perceived quality substantially. Therefore, when stable transmission in either HSDPA or WiFi can be guaranteed, switching between those networks is not recommended. Moreover, if a handover can be foreseen, proactive scheduling of the entire transmission in a lower link quality network allows to maximize the user experience. Otherwise, a degradation due to a network handover is experienced, which, however, can be masked by a low encoding bit rate or robust video coding. As a matter of fact, the video codec must be selected to match the networking circumstances of a video call. When a stable connection cannot be guaranteed, the use of the MPEG-4 codec is recommended, but if a reliable network is available, H.264 should be used to maximize the QoE. Moreover, the overall quality of a video call can be also improved by maximizing the encoding bit rate, which is of particular relevance for codecs such as MPEG-4 that are not optimized for low transmission bit rates. Therefore, if the connection throughput is limited, H.264 should be preferred over MPEG-4. Finally, the Quality of Experience during a video call can be additionally impoved through high quality of the speech signal, which is of particular relevance when the video quality is high as well.

In contrast to the results of the previous test, switching between codecs or bit rates in a video call is always perceived according to the logical expectations. A switch to H.264 or to a higher bit rate always improves the Quality of Experience, but it should be considered that an improvement is less perceived than a quality degradation that, in turn, leads to quality ratings that are close to that of constantly low quality. Therefore, oscillation of the provided quality should be avoided any time. If it can be foreseen that the video quality will have to be degraded, constant provisioning of low quality enables to reduce the service management effort at no substantial costs of the perceived quality. Moreover, switching between narrowband and wideband speech during an ongoing video call should be avoided, and alternatively constant narrowband quality should be provided. However, if the transmission must be adapted, it is recommended to change the video configuration first, before the bandwidth of the speech signal is switched. Low video quality reduces the sensitivity of video telephony users to degradations that appear in the audio channel. Moreover, quality adaptation is also advantageous when transmission degradations occur and once again a changeover to a more robust codec turned out to be the adaptation technique that always improved the Quality of Experience. However, if packet loss is high, network handover is the most recommended technique of transmission adaptation, which is of particular relevance when a mobile user is leaving a WiFi area. It was also shown that the reduction of the video encoding bit rate is not an effective technique of quality adaptation when link-layer impairments occur. Finally, if connection degradations and the need of transmission adaptation can be foreseen, proactive use of a robust video codec or permanent streaming in a loss-free but lower-quality network enable to deliver higher Quality of Experience than any other of the dynamic adaptation methods that were considered in this section.

5.2.3 Quality of Emulated Video Calls

The user tests that were so far evaluated in this chapter were focused on the quality perception in a pure listening and viewing environment only. To link those results to the Quality of Experience in a situation that is close to a real video call, Test V3 was carried out. The analysis of this test was structured in three groups. And so, user perception of constant quality of a video call is presented as first in this section. Next, the Quality of Experience due to user mobility is addressed. Finally, the impact of the temporal position of a quality change on the overall experience of a video call is evaluated as last.

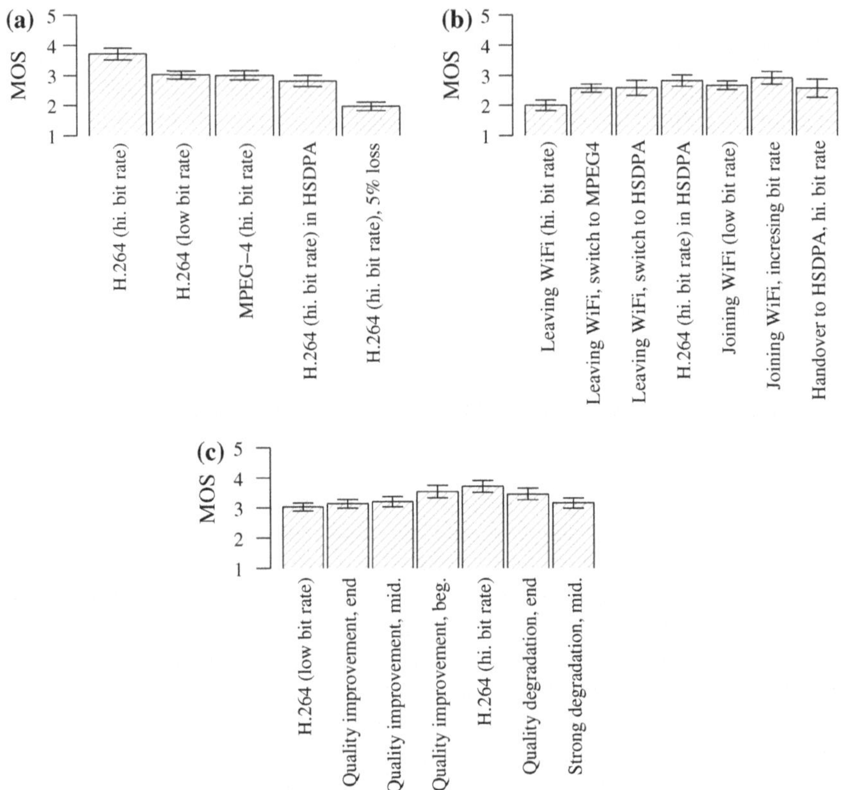

Fig. 5.4 User judgements of quality in Test V3. MOS_{av}, and 95 % CI. Selected groups to address user perception of: **a** Stable transmission, **b** Mobility, **c** Temporal position of a quality change

The collected results (cf. Fig. 5.4) confirmed that when a video call was performed in an unimpaired network, H.264 at 1536 kbit/s provided the highest Quality of Experience. However, even if this profile offered the highest quality in Test V3, its MOS score was lower than "good". Moreover, when the MPEG-4 codec was used, the Quality of Experience was substantially degraded. The quality judgement in this case was similar to that collected for a video call that was encoded at the bit rate of 256 kbit/s with H.264. In addition, all of the above profiles were rated lower than the corresponding samples in Test V2, which revealed a general trend that the quality of a video call was rated lower than that of a short transmission extract. Finally, the Quality of Experience was also substantially degraded, when the reference transmission with H.264 at 1536 kbit/s was performed in a HSDPA network. However, this observation is related to the fact that the HSDPA access was provided via a public hot spot without any resources guarantee for the project. Thus, network related degradations that resulted of an overloaded hot spot occurred.

As a result, the use of H.264 at 1536 kbit/s enables to maximize the call quality experience under stable networking conditions. However, in opposite to the results of Test V2, the quality in both, WiFi and HSDPA, was considerably different, which shows that also the load of a public hot spot must be considered in the mobility management to assure high Quality of Experience. In fact, any change of the reference streaming configuration in any of the available dimensions, such as use of a different network, codec, or a lower encoding bit rate causes a substantial loss of quality. However, the knowledge that H.264 is a more efficient coding scheme than MPEG-4 helps to optimize the streaming resources at no cost of perceived quality.

When the quality of a video call in Test V3 was affected by user mobility, increasing packet loss degraded the Quality of Experience substantially. The quality impression in such a situation was similar to that of a call that was constantly affected by 5 % of packet loss. When the effect of an increasing packet loss was countered by a handover to a loss-free HSDPA network or by a changeover to the MPEG-4 codec, the Quality of Experience was always substantially improved. However, in any case, the overall quality was judged lower than that of constant transmission in HSDPA. Therefore, when packet loss and the need of quality adaptation can be foreseen, it is advantageous to proactively schedule the entire transmission in a HSDPA network, as no perceptual profit of the initially unimpaired WiFi is taken, and the overhead of mobility management can be reduced. This was additionally confirmed by the quality judgement of a quality profile that included a network handover from WiFi to HSDPA in the middle of a call, and that was not degraded by packet loss. The quality of this profile was rated below that of constant transmission in HSDPA as well.

In turn, when the packet loss rate was decreasing during a video call, the end quality of such a connection was rated closer, although substantially lower, to that of a profile that was not affected by packet loss at all. Moreover, if the encoding bit rate was increased in the middle of a call to boost the user experience while entering a WiFi area, the Quality of Experience was improved, but the perceived gain was not substantial. All these finding compared with the observations from the previous paragraph once again confirm that a quality degradation is remarkably stronger perceived than a quality improvement.

As a result, all of the analysed adaptation techniques helped to improve the Quality of Experience during a video call. However, the improvement was only substantial, when a technique was used to counterbalance the effect of increasing packet loss. Nevertheless, when packet loss and the need of quality adaptation can be foreseen, it is advantageous to schedule the entire transmission in a wide-coverage, but unimpaired, network. Such a proactive decision helps to maximize the experience of video call quality, and at the same time it allows to reduce the overhead of quality management.

The last research aspect that was addressed in Test V3 was the relationship of the overall Quality of Experience with the temporal position when the quality is changed within a video call. When the encoding bit rate of H.264 was increased from 256 to 1536 kbit/s in the first, middle, or in the last segment of an emulated call, the overall Quality of Experience was always improved. The earlier in a video call the adaptation was performed, the stronger was the perceived effect, but the improvement was substantial only, if the bit rate was increased at the beginning of a call. In any other case, the quality judgements were similar to each other. In turn, when the video encoding bit rate was decreased at the end of a video call, the observed effect was once again stronger than that of a quality improvement at the same position in a call. Moreover, because a degradation of quality at the end of a call was considered as the worst case of a quality change in this analysis, but the observed effect was not substantial, it was discovered that not the position when the quality is changed but the duration of low quality transmission is the deciding aspect of quality experience. Finally, when the middle segment of an emulated call was degraded by 5 % of packet loss, the end quality judgement was substantially lower than that of an unimpaired profile. This observation confirms that not only the position and duration of an impairment, but also its strength is essential for the Quality of Experience.

As a result, the above analysis confirmed that the temporal position when the quality is changed in a video call affects the Quality of Experience. Even though an improvement of quality during a video call is always perceived by the users, its effect is substantial when the quality is adapted early in a call only. A degradation of the quality within of a video call is perceived stronger than a quality improvement, but rather than the position when the quality is degraded the duration of the exposure to a low quality signal is of particular relevance. Moreover, the strength of an impairment also influences the quality perception and thus peak degradations in a call might substantially reduce the overall Quality of Experience. Therefore, when such degradations are expected, constant use of a robust video codec is recommended, at no loss of the perceived call quality.

The above presented analysis enabled to extend the set of quality management recommendations that were derived in this work. It was confirmed that the quality perception in wireless networks depends on the position, duration, and strength of quality degradations that might occur in a video call. In this context, the quality of an available network connection and the load of a public access point must be considered to assure high Quality of Experience while roaming across heterogeneous networks. To this end, the video codec should be matched for the networking conditions to maximize the Quality of Experience. When the networking circumstances are unimpaired, the video calls that are transmitted using H.264 at 1536 kbit/s achieved the best quality ratings, but any change of this configuration degraded the Quality of Experience substantially. However, if any, even short, link layer degradations are expected, a robust video codec, such as MPEG-4 in this work, is recommended at no loss of the overall quality that is perceived by the users. Moreover, use of quality adaptation techniques is always advantageous. However, the quality improvement is substantial only when the adaptation is applied to counterbalance the visual effect of packet loss, or if it is used early in a call. Nevertheless, if a quality degradation due to packet loss can be foreseen, proactive scheduling of the entire video call in a lower quality network, such as HSDPA, helps to maximize the Quality of Experience. Therefore, a trade-off between the effort of quality management and the expected perceptual effect should be optimized.

5.3 Instrumental Quality Prediction

As already mentioned in the previous chapter, the success of telephony offerings requires reliable tools to instrumentally assess the quality of the provided service. Even thought all that is also true for video telephony, most of the efforts in developing quality models were focused on the speech telephony until now. The majority of video quality models that are applicable for estimation of video telephony quality have neither been strictly designed for this type of service nor for estimation of time-varying quality that might occur in mobile networks. Therefore, the models are tested outside of their original use case in this work. Four types of models are addressed. Speech, video, audiovisual integration models, as well models for time-integration of quality during a call—so-called "call quality models". The last type of models was developed for speech telephony only and therefore it is additionally validated for applicability in video telephony systems in this work.

The analysis that is presented in this section was based on a comparison of the user judgements that were collected during the user tests that were presented in this chapter with quality estimates that were provided by the addressed models. Both quality metrics were compared per condition on the 5-point MOS scale without any additional intermediate mapping of the input data to avoid any tuning of the results and so to provide knowledge that is relevant for real systems. This approach is consistent with that already applied in the previous chapter. In this context, the accuracy of the models was analysed based on the Pearson correlation and the Root Mean Square Error (RMSE) of the quality metrics, and additionally the inaccuracy of the models was compared according to the methodology that was proposed in [12], as RMSE is considered as a decisive accuracy factor in this study. The reasons of estimation error were discussed and the significant problems were explored using Analysis of Variance (ANOVA) that was based on a linear model with the estimation error per test condition as the dependent variable and with the speech and video codec, network, and packet loss as the independent variables. Doing so, switching of the transmission parameters was modeled as additional levels of the addressed factors. The normality of the error population was examined visually prior to analysis.

5.3.1 Accuracy of Sample Quality Models

Video Quality Models

The perceived quality of short video samples that are typically few seconds long can be assessed, similarly to the quality of speech samples, with various prediction models that differ in their purpose, range of degradations they can cope with, and the functional principle (cf. Sect. 2.2.2). Three of those models that are recommended by ITU-T and that were designed for different purposes were analysed in this work:

- ITU-T Rec. J.144 (aka "VQM"),
- ITU-T Rec. J.341 (aka "VQUAD"),

- Video model from ITU-T Rec. G.1070, further referred to as G.1070.

All of these models were applied to estimate the quality of video samples that were used in Test V1 and V2. The G.1070 model was utilized according to the standard recommendation with an additional set of codec coefficients (cf. Table A.16) that were derived in a separate test and that were proposed to ITU-T in [13]. Moreover, if a sample was composed of two blocks of different quality levels, an arithmetic mean was computed on the MOS scale. In turn, the VQM estimates that were originally expressed on the models own scale ($VQM \in [0; 1]$) have been linearly mapped on the 5-point MOS scale. This transformation did not affect the correlation of the results but helped to compare the estimation error between the analysed models.

The collected results (cf. Table 5.3) revealed that the quality estimates that were provided by G.1070 and J.341 were the most correlated with the user judgements. However, the top ranking of the models in terms of the prediction error was swapped in both tests. According to the purpose of the model, assessment of the quality of various content types in Test V1 was more accurate when J.341 was used, but the quality of the signal that is typical for video telephony in Test V2 was estimated best by G.1070. The quality scores that were obtained with the J.144 model suffered from the lowest correlation with the user ratings and from the largest prediction error in both of the tests. This does not come at a surprise, as the J.144 model was designed for estimation of coding artifacts only and other types of degradations were not considered in the design of this model.

A comparison of the RMSE values that were achieved by all of the tested video models disclosed that the prediction error of all of the methods was lower in Test V1 than in Test V2. This effect might be related to the extended set of the test material that included additional content types in the first test, or to a cross-influence of audio and video that was addressed in Test V2 only. However, more dedicated tests are required to address this issue. In this context, the estimation errors that were introduced by J.341 and G.1070 were statistically equal and both of them were significantly smaller than that of J.144 in both of the tests. However, it is emphasised that the quality estimates that were provided by J.144 have been linearly mapped on the common scale.

A direct comparison of the individual quality estimates with the corresponding user judgements (cf. Figs. 5.5 and 5.6) showed that the video quality scale was used in a similar range by the users and models in both of the tests. All this despite the presence of audio in Test V2, which is in line with the results of the user tests that

Table 5.3 Accuracy of video quality estimates in Test V1 and V2

	Test V1		Test V2	
Name	Pearson	RMSE	Pearson	RMSE
G.1070	0.95	0.35	0.94	0.39
J.341	0.89	0.33	0.89	0.40
J.144	0.76	0.75	0.83	1.16

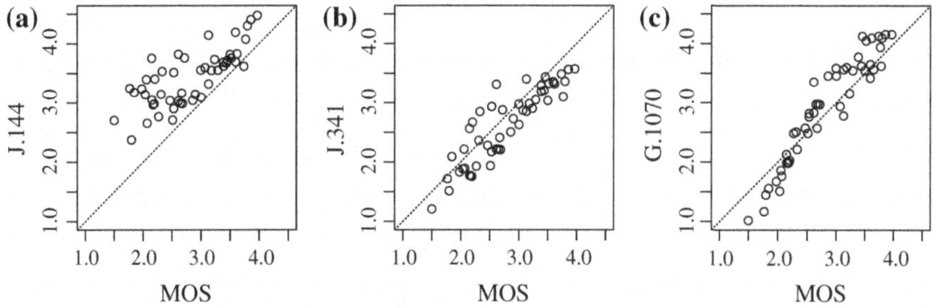

Fig. 5.5 Comparison of video quality estimates and user judgements (MOS_v) in Test V1. **a** J.144, **b** J.341, **c** G.1070

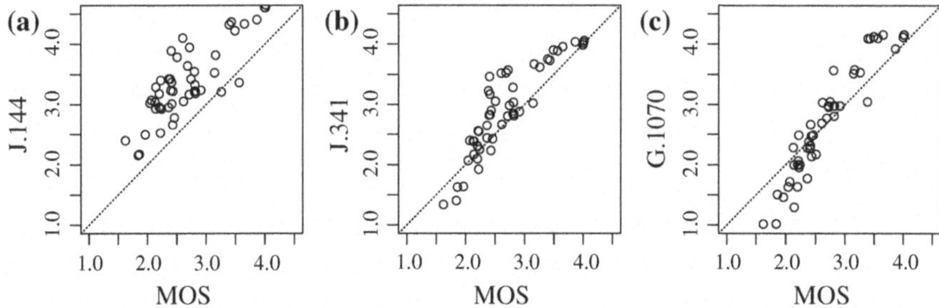

Fig. 5.6 Comparison of video quality estimates and user judgements ($MOS_{v|av}$) in Test V2. **a** J.144, **b** J.341, **c** G.1070

were presented in the previous section. Moreover, also the general trends of quality estimation were similar in both tests. The predictions that were provided by G.1070 and J.341 were close to user opinion, although certain differences between the outputs of those models were observed. The quality estimates that were provided by G.1070 were overestimated and underestimated in the upper and lower parts of the scale, respectively. This effect discloses that the edge conditions that either represented high video quality or that were affected by the strongest degradations of the test were estimated worst. These trends were stronger in Test V2. Therefore, a closer look to the group of the conditions that were overestimated in that test revealed that when high quality video was provided, any discrepancy in the high bit rate transmission from the reference configuration that was defined as H.264 at 1536 kbit/s in WiFi resulted in a quality overestimation. This was for example the case when high bit rate video was transmitted in HSDPA, when MPEG-4 was used, or when conditions with the highest encoding bit rate of the test were affected by either network or codec switch. This indicates that the instability of transmission circumstances when high quality video is provided must be more carefully modeled in G.1070. However, those conditions were not the most overestimated ones in this test. The largest overestimation error of G.1070 was observed when the quality of a stimulus was rapidly degraded due to a reduction of the encoding bit rate in the middle of a sample. This is a prove that pure averaging of quality scores does not allow to accurately assess the quality of time-varying transmission, and that the results from the user tests that showed that a degradation of video quality is stronger perceived than its improvement must be additionally modeled. This can be realized with approaches such as that presented in [14]. In turn, a closer look to the other group of conditions that were underestimated by the G.1070 model showed that samples that were constantly or partially affected by the highest packet loss value of the test were underestimated most, which reveals that the model was too pessimistic when substantial quality degradation occurred. In this case, however, a new set of codec coefficients might reduce the inaccuracy.

In turn, the relationship of quality estimates that were provided by the J.341 model and user judgements was slightly different between the conducted tests. In Test V1, this model predominantly underestimated the quality and this behaviour was approx. constant across the entire scale. However, the underestimation was lower in Test V2, where a peak overestimation of conditions that were judged by the users in the range between approx. 2.3 and 2.9 MOS was disclosed. This group included predominantly conditions that were affected by the lowest packet loss rate of the test, including conditions in which the video quality was adapted. This finding showed that once again packet loss was one of the main reasons of quality estimation error, although the strongest degradations of the test were well modeled by J.341. Finally, the quality predictions that were obtained with the J.144 model were mostly overestimated in both of the tests. However, it is once again emphasised that those results were linearly mapped to the common scale.

The main reasons of the above described estimation errors were confirmed through a statistical analysis (cf. Table 5.4). The prediction accuracy of the two top performing models, G.1070 and J.341, was predominantly affected by packet loss in both of the tests. Moreover, in Test V1, the accuracy of G.1070 was reduced due to the video codec, and that of J.341 was slightly degraded due to the video bit rate. In this test also the network technology slightly reduced the accuracy of quality prediction that were provided by both models. However, in Test V2 the encoding bit rate was the only factor that was slightly contributing to the estimation error of both models, apart of packet loss. These results show that the particularities of user perception in heterogeneous wireless networks that were discovered in the previous section should

Table 5.4 Statistical contribution of transmission parameters to the error in video quality prediction in Test V1 and V2

	Test V1						Test V2					
	J.144		J.341		G.1070		J.144		J.341		G.1070	
Factor	F	p	F	p	F	p	F	p	F	p	F	p
Packet loss	11.17	2.4e-06	7.75	5.7e-05	8.59	2.4e-05	4.38	0.0024	3.73	0.0063	10.65	1.7e-06
Network	3.43	0.021	2.37	0.076	2.71	0.050	0.80	0.4594	0.01	0.9911	0.37	0.695
Video bit rate	1.41	0.236	2.28	0.051	1.03	0.440	3.00	0.0451	2.33	0.0930	2.78	0.057
Video codec	1.32	0.288	2.18	0.113	4.13	0.015	6.48	0.0015	0.58	0.6302	0.72	0.549
Speech codec	NA	NA	NA	NA	NA	NA	0.21	0.8857	0.93	0.4353	0.20	0.893

be considered in quality prediction models that are applied in mobile networks. In particular, it was shown that quality degradations that are not expressed by the configured packet loss, for example the instability of a network, must be separately included in the parametric modeling using G.1070, which is of particular relevance when the quality of high bit rate video is assessed. The statistical analysis also confirmed that the estimates that were provided by the J.144 model were inaccurate due to packet loss and the network in Test V1 in this order or relevance, which is an expected result, as the model was not designed for this type of quality degradations. However, this trend was different in Test V2, where the video codec, packet loss, and the video bit rate were the factors that were significantly contributing to the error of the linearly mapped quality estimates that were provided by this model.

As a result of the above analysis, the accuracy of the addressed video quality prediction models in heterogeneous wireless networks primarily depends on the purpose of a model. The quality of various content types was well predicted by the J.341 model, but in case video telephony was addressed, G.1070 provided the most accurate quality estimates. The J.144 model with a linear mapping of the results has not produced satisfying results in the context of mobile networks, which is also not the purpose of the model. Moreover, although the top performing models were developed to capture transmission-related degradations, packet loss was the main reason of the estimation error in all of the conducted tests. Moreover, the instability of networking circumstances must be more carefully modeled in G.1070, which is of particular relevance when the quality of high bit rate video should be assessed.

Finally, G.1070 was not further extended in this work to enable this model for assessment of time-varying video quality in mobile networks. Although it was shown that plain averaging of different quality levels of a video telephony sample in Test V2 did not provide satisfying results, only four of the 50 test conditions were substantially affected by this type of prediction inaccuracy. Therefore, no significant results of such a modeling were expected. Moreover, it was proven that in the current shape of the model the estimation error was related to packet loss most, which included bursty losses due to a network handover. Accordingly, those are the aspects of the model that should be improved first, or a new set of test data must be generated that is focused on time-varying quality due to switching between video codecs and their encoding bit rates.

Speech Quality Models

The accuracy of speech quality models in the context of audiovisual transmission was evaluated next. The analysis followed the same methodology to that already presented in the previous chapter, and also the same set of quality prediction models was addressed in the analysis:

- ITU-T Rec. P.863 (aka "POLQA"),
- ITU-T Rec. P.862.2 (aka "WB-PESQ"),
- ITU-T Rec. G.107.1 (aka "E-model").,

All of the models were evaluated based on the data from Test V2. In this context, the E-model was applied according to the extension that was presented in Sect. 4.3.2.

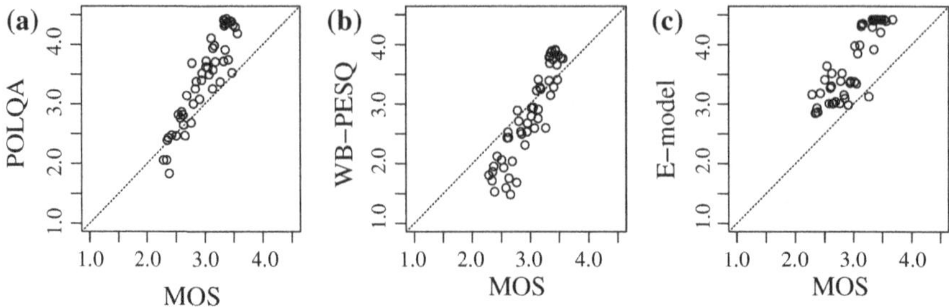

Fig. 5.7 Comparison of speech quality estimates and user judgements in Test V2. **a** POLQA, **b** WB-PESQ, **c** E-model

If a sample was composed of two blocks of different quality levels, a mean quality estimate was computed. If the speech codec was switched between wideband and narrowband quality, the bandwidth switching impairment factor I_{sw} was used to correct the error of quality estimation.

The collected results (cf. Table 5.5) revealed that the quality estimates that were obtained with WB-PESQ were the most accurate in terms of the prediction error, although the output of POLQA was slightly more correlated with the user opinion. The quality estimates that were provided by the E-model were the most inaccurate ones in terms of both, correlation with user judgement and the estimation error, in the entire test. An additional comparison of the RMSE vales between the models revealed that all of the values were statistically different from each other. This made WB-PESQ the most accurate full-reference model of the test, and simultaneously the advantage of both signal-based models over the parametric E-model was statistically confirmed. This findings are similar to the observations that were made in the speech telephony context of the previous chapter. However, the results of both tests are not fully consistent with each other. In contrast to the previous analysis, the accuracy ranking of the signal-based models was swapped when the video signal was provided. Some reasons of this problem are related to the interaction between speech and video quality which is explained later in this section, although it should be noted that none of the signal-based speech quality models was designed for the application in a video telephony system.

A direct comparison of the estimates and user judgements of speech quality in Test V2 (cf. Fig. 5.7) revealed that the user scores were distributed in a narrower range on the quality scale than the values that were provided by the models. In this context, even if quality estimates that were produced by both of the signal-based models were correlated with user opinion, the results were biased. The WB-PESQ

Table 5.5 Accuracy of speech quality estimates in Test V2

Name	Pearson	RMSE
WB-PESQ	0.91	0.47
POLQA	0.93	0.61
E-model	0.86	0.81

model predominantly underestimated, and POLQA predominantly overestimated the speech quality. The wideband E-model systematically overestimated the quality of most of the test conditions. All those general trends were in line with those observed in the previous chapter, although the inaccuracy was amplified in the audiovisual context of a video call.

A closer look to the quality estimates that were provided by WB-PESQ and POLQA disclosed an interaction between the estimation error and the video quality of a sample. Pure wideband speech that was provided when the video quality was limited due to a low encoding bit rate, in particular when MPEG-4 was used, was the most overestimated case that was observed for both of the signal-based models. Apparently, low video quality degraded the experience of speech quality during a video call. Also the reasons of quality underestimation were similar for both models, although certain differences were be observed. Both, WB-PESQ and POLQA, substantially underestimated the speech quality, when a changeover between wideband and narrowband speech occured. Those were consistently the four most underestimated conditions of WB-PESQ, which means that in a video call the impact of a changeover between wideband and narrowband speech is less critical in a speech-only context. A substantial underestimation of speech quality was also observed when a sample was degraded by packet loss and high bit rate video was provided. In this context, WB-PESQ was more inaccurate when a wideband codec was used and stronger packet loss was observed, whereas POLQA was in particular inaccurate when narrowband speech quality was provided.[1] Apparently, users are less sensitive to distroted speech in an audiovisual context, even if the quality of the video signal is degraded.

The presented results are partially in line with the findings that were discovered in the previous chapter. Consistently accross both studies, WB-PESQ was a more pesimistic model that in particular punished a switch between narrowband and wideband codecs during an ongoing session. An inverse behaviour between the tests was observed when POLQA was used. While this model in general positively assessed the quality of narrowband speech in a VoIP test, those ratings were more negative in context of the video telephony that was addressed in this chapter. The E-model provided systematically overestimated results in both studies.

The main reasons of the estimation error were confirmed through a statistical analysis (cf. Table 5.6). The accuracy of all the addressed models was primarly affected by the speech codec and packet loss. Additionally, the accuracy of all three models was significantly degraded by the video encoding bit rate that was the third most contributing factor to the estimation error of the signal-based models, and the fourth most affecting factor of the E-model prediction. In the last case, more than the video bit rate the network that, however, also significantly degraded the accuracy of the WB-PESQ model, and slightly significantly that of POLQA, reduced the prediction accuracy. Further investigation of the last observation revealed that the test material that introduced a network handover from WiFi to HSDPA was affected

[1] POLQA includes an extra narrowband mode of operation that is dedicated to narowband speech-only. This mode, however, is not applicable in a multi band environment.

by bursty packet loss that was related to the video transmission bit rate. The higher the video bit rate was, the stronger was the degradation. This quality loss turned out to be not well estimated by the models, which was in particular problematic in case of the parametric quality prediction.

As a result of the above analysis, in the audiovisual context of a video call user judgements of the speech quality are expressed in a narrower range of the scale than the quality predictions that are provided by regular speech quality models. Therefore, although the quality estimates are well correlated with the user judgements, the values are biased. This effect indicates that when the speech quality of a video call should be measured, the quality estimates that are provided by regular prediction models must be transformed to express the speech quality in the audiovisual context of a video call. Moreover, signal-based models once again turned out to predict the speech quality more accurately than the parametric E-model. However, the reasons of the estimation error were similar for all models, being packet loss and speech codec, in particular including a changeover between wideband and narrowband speech, the two main reasons of quality estimation error. Moreover, the network, in particular a network handover, in relationship with the video encoding bit rate were causing signal interruptions, which was another reason of quality assessment error that was in particular critical in case of the parametric quality prediction.

Audiovisual Integration

When the quality of the speech and video signal of a sample is known, both of the values can be integrated to estimate its audiovisual quality. For this purpose, an integration model is recommended in ITU-T Rec. G.1070, and the accuracy of this method is analysed in this section.

$$MOS_{int|av} = 0.147 MOS_{a|av} + 0.262 MOS_{v|av} + 0.09 MOS_{a|av} MOS_{v|av} + 0.837 \tag{5.1}$$

The integration function in this work was used in conjunction with modelling coefficients that were derived in [13] (cf. Eq. 5.1). In this form the model was applied to predict the quality of audiovisual stimuli from Test V2. In the first stage of this evaluation, user judgements of the speech and video quality of the test samples were

Table 5.6 Statistical contribution of transmission parameters to the error in speech quality prediction in Test V2

	POLQA		WB-PESQ		E-model	
Factor	F	p	F	p	F	p
Packet loss	8.05	2.4e-05	11.49	7.9e-07	22.89	2.7e-10
Network	2.56	0.0928	4.15	0.02502	10.83	0.00026
Video bit rate	4.60	0.0087	7.31	0.00072	4.82	0.00702
Video codec	2.03	0.1289	6.48	0.00150	1.83	0.16101
Speech codec	16.83	9.7e-07	36.81	1.7e-10	14.06	5.1e-06

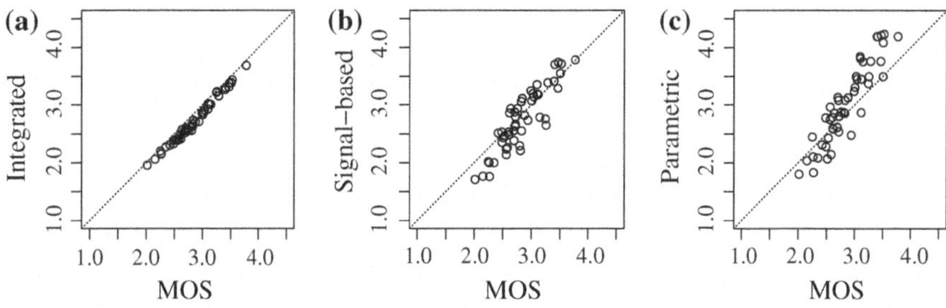

Fig. 5.8 Comparison of estimates and judgements of audiovisual quality in Test V2. **a** Integrated user judgements of audio and video quality, **b** Integrated Signal-based estimates, **c** Integrated parametric estimates

integrated, and the obtained estimates were compared with the ratings of the audio-visual quality that were collected during the test. Afterwards, the integration model was tested in conjunction with quality predictions that were provided by the quality models that were addressed previously in this section. Doing so, two combinations of the top performing models, a parametric and a signal-based approach, were created in this analysis. The parametric method was compliant with ITU-T Rec. G.1070 and thus the quality estimates that were provided by the video model from this recommendation and by the extended E-model were integrated. The signal-based approach was based on the quality estimates that were provided by J.341 and WB-PESQ that were identified as the most accurate signal-based models in context of audiovisual transmission in heterogeneous wireless networks.

The collected results (cf. Table 5.7) showed that when the user judgements of speech and video quality were integrated, the estimates of audiovisual quality were close to user opinion. As a matter of fact, this was the most accurate model in this work, and ANOVA applied on these results has not revealed any significant reasons of the estimation error. However, the accuracy of the audiovisual estimates substantially decreased when predicted values were used as a modelling input. In this case, the signal-based approach generated more accurate results than the parametric approach, which was expressed as a significantly lower RMSE value in the former case.

A direct comparison of the fully estimated results with the user judgements of audiovisual quality (cf. Fig. 5.8) revealed that the integration process compressed the quality scores in the middle of the quality scale, which was in line with the user opinion. Moreover, the peak of quality overestimation introduced by the signal-based J.341 model was reduced through the integration with the generally underestimated

Table 5.7 Accuracy of audiovisual quality integration in Test V2

Name	Pearson	RMSE
User	0.99	0.14
Signal-based	0.89	0.26
Parametric	0.93	0.39

quality scores that were provided by WB-PESQ. In case of the parametric approach, the estimates of audiovisual quality were shifted to the upper range of the scale, which resulted in a particular overestimation of quality in the upper range of the MOS scale.

As a result of the above analysis, the audiovisual integration model provides quality estimates that are close to the user opinion. However, when predicted quality scores are integrated instead of user judgements of speech and video quality, errors that affect both groups of values are projected in the estimates of the audiovisual quality. This fact enables to improve the accuracy of quality prediction when the models are carefully selected.

5.3.2 Accuracy of Call Quality Models

All of the quality models that were evaluated so far in this chapter were designed to predict the quality of short transmission samples only. However, when the quality of a stimulus that has a structure of a typical phone call should be measured, the use of so called "call quality models" is recommended. Four of those models were addressed in this section and their selection was in line with that already introduced in the previous chapter:

- model proposed in ETSI 102 506 v.1.2.1 (aka "ETSI model"),
- model proposed in [15] by Weiss et al. (aka "Weiss model"),
- model proposed in [16] by Rosenbluth (aka "Rosenbluth model"),
- baseline averaging of the sample quality (aka "Mean model").

All of these models were originally designed to assess the quality of a speech call only, which is why all of them are tested outside of their original use case in this section. To estimate the quality of the video calls from Test V3, user judgements of audiovisual quality that were collected in Test V2 were used as input. The models were not adapted to the audiovisual context in any way, so they were used as recommended by their authors; in focus of the presented investigation are the actual call quality models. The baseline approach of quality averaging is provided for initial comparison only.

The collected results (cf. Table 5.8) revealed that the quality estimates that were provided by all of the models were well correlated with user opinion and that only subtle differences between the compared tools were observed. This observation proves

Table 5.8 Accuracy of video call quality estimates in Test V3

Name	Pearson	RMSE
Mean	0.94	0.29
ETSI	0.95	0.29
Weiss	0.95	0.22
Rosenbluth	0.96	0.23

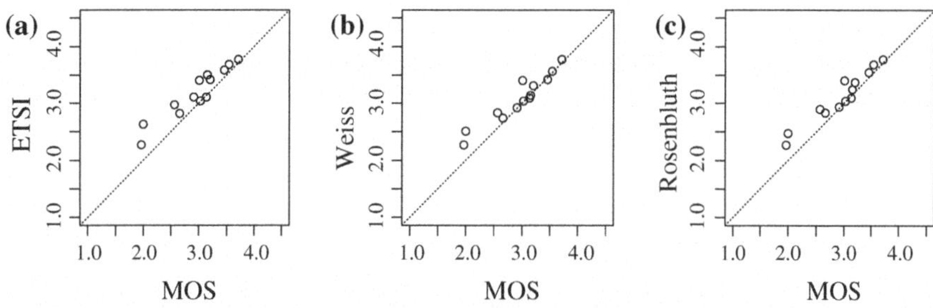

Fig. 5.9 Comparison of estimates and user judgements of video call quality in Test V3. **a** ETSI, **b** Weiss, **c** Rosenbluth

that the call quality models in an unmodified form can be applied to assess the quality of emulated video calls. In this context, the Weiss model provided the most accurate quality estimates, which was indicated by the lowest prediction error in the entire test. A slightly less accurate result was achieved by the Rosenbluth that generated quality estimates that were, however, slightly more correlated with the user opinion. Both of the models, Weiss and Rosenbluth, were more accurate than the model that is proposed by ETSI. The output of this method was always affected by the largest prediction error, and the inaccuracy in this case was comparable with that of plain averaging of the quality scores that belong to the individual segments of a video call. Plain averaging also resulted in the lowest correlation of the quality estimates with the user opinion in the entire test. Despite all of the aforementioned differences between the models, a comparison of the RMSE values confirmed that their accuracy was statistically equal.

Further analysis of the call quality estimates (cf. Fig. 5.9) revealed that overestimation, i.e. too optimistic quality assessment, was the main reason of the estimation error, which was observed for all of the analysed models. The effect was particularly noticeable when a quality profile was substantially degraded. For example, when a high bit rate stream of H.264 at 1.5 Mbit/s was affected by an increasing packet loss of 0, 3, 3, 5, 5 %, for the consecutive samples of a call, respectively, the overall quality of a video call was overestimated most. The estimation error of 0.57, 0.45, 0.41 MOS that was introduced by the ETSI, Weiss, and Rosenbluth model, respectively, was always significant ($p < 0.05$). A similar effect was observed when the same pattern of degradations was experienced, but the connection quality was adapted through a switch to MPEG-4 in the middle of a call. Apparently, there are two reasons of this inaccuracy. First, as already presented in this chapter long exposure to bad quality is perceived worse than a short sample of the same quality degradation that, however, is used for the modeling. Second, rapidly decreasing quality is not well covered by the call quality models, which, however, is not related to the presence of video. A similar observation was already made when the quality of super-wideband calls in Test S3a was analysed. The last profile that was commonly overestimated by all of the models introduced constant quality degradation by 5 % of packet loss, which corresponds to the first reason of estimation error that was mentioned above.

In this context, no more significant errors were observed when the Weiss model was used. However, when the Rosenbluth model was applied, a profile where the video encoding bit rate was increased from 256 to 1.5 Mbit/s in the middle of a call was significantly overestimated, and the same profile, as well as a profile whose middle sample was degraded by 5 % of packet loss were incorrectly assessed by the ETSI model. The last observation confirms that taking the worst quality sample of a video call in consideration, which is the case of the Weiss model, improves the accuracy of quality prediction.

The presented evaluation proved that the currently existing call quality models enable to reliably assess the quality of emulated video calls. Moreover, despite the audiovisual context of the analysis in this section, the achieved results were in line with that already discovered when the quality of speech calls in the previous chapter was assessed. The common weakness of all the call quality models was quality overestimation, which was particularly observed when either the transmission quality was substantially degraded, which is related to the fact that long exposure to bad quality is perceived worse than a short sample of the same quality degradation, or when the quality of a call was rapidly decreasing. Therefore, the models must be improved for this use case. Moreover, additional weighting of bad transmission periods of a video call turned out to improve the accuracy of quality prediction. This proves that also in case of video telephony the impact of a bad quality sample is relevant for the modeling.

5.3.3 Instrumental Modeling of Video Call Quality

After the accuracy of the individual prediction models that may be applied for assessment of video call quality in heterogeneous wireless networks was evaluated, the combined use of those methods was analysed. This is the subject of this section.

Four types of models, one for speech, one for video, one for audiovisual, and one for call quality estimation were used together to fully instrumentally assess the quality of a video call. Due to a large number of possible model bundles, consistently to the methodology that was already introduced in the previous chapter, only two of the possible approaches, a fully parametric and a signal-based method, were analysed in this section. The parametric approach was based on ITU-T Rec. G.1070 and on the Weisss call quality model that was selected as the most accurate method in this work. Accordingly, the video model from G.1070 and the E-model (including the enhancement that was proposed in Sect. 4.3.2) were applied to estimate the quality of a sample, and both of the quality metrics were integrated using the audiovisual integration model that is recommended in ITU-T Rec. G.1070, and the same that was analysed in Sect. 5.3.1. The audiovisual estimates were then used as an input to the Weiss model to assess the overall quality of an emulated video call. In turn, the signal-based approach was based on a set of signal-based models, WB-PESQ and J.341, that were selected as the most accurate full-reference models in this chapter.

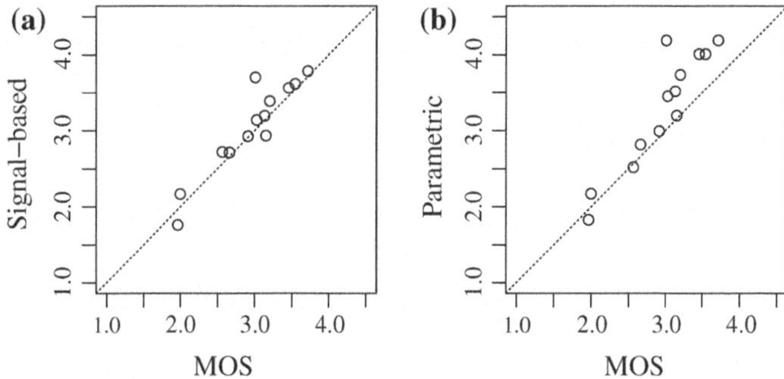

Fig. 5.10 Comparison of instrumental estimates and user judgements of video call quality in Test V3. **a** Signal-based, **b** Parametric estimates

The estimation error of both approaches was evaluated in analogy to the methodology that was presented in the previous section. The estimates of the sample quality from Test V2 were integrated and applied to model the quality of the emulated video calls in Test V3. The predictions and user judgements of the audiovisual call quality were then compared.

The quality estimates that were collected with both of the analysed approaches were less accurate than when user judgements were used as an input to predict the call quality (cf. Table 5.9 and Fig. 5.10). However, the estimates that were collected with the signal-based method were close to the results in the previous section, excluding the exceptional case of constant MPEG-4 transmission that was substantially overestimated. As a result, the combination of the prediction errors that were introduced by the individual models of the signal-based approach enabled high accuracy of video call quality prediction. In particular, the audiovisual integration of the quality estimates that were provided by J.341 and by WB-PESQ resulted in a shift of the integrated values towards the bottom of the scale, which enabled to compensate for the overestimation that was observed when the Weiss model was used.

In turn, when the parametric approach of video call quality assessment was applied, a biased overestimation of conditions that were rated in the upper range of the quality scale was discovered. Quality profiles that included a large portion of H.264 transmission at 1536 kbit/s were affected by this effect most. This included a profile that represents constant transmission with this configuration, improvement of quality at the beginning or in the middle of a call, and a profile that was degraded in the last sample. Apparently, this finding is related to the overestimation of high quality video by G.1070 and an additional amplification of this effect through audio-

Table 5.9 Accuracy of fully-instrumental call quality estimation in Test V3

Name	Pearson	RMSE
Parametric	0.93	0.47
Signal-based	0.93	0.23
User	0.95	0.22

visual integration of the results with the as well overestimated quality scores that were provided by the E-model.

The study of fully instrumental assessment of video call quality confirmed that careful selection of quality models in heterogeneous networks is required, and in particular the additivity of the prediction error at different stages of the modeling process must be considered. A sophisticated combination of those methods, such as the presented signal-based approach, helps to reduce the overall estimation error. However, if models that cannot handle certain degradation types are used together, the error of quality prediction might be amplified. Both, the parametric and signal-based approach, are considered in this work as relevant for quality monitoring in real telecommunication systems. They essentially differ in the type of input information that is used to predict the quality of a video call, which defines different purposes of their use. The parametric approach does not require any access to the physical signal, and so it can be used for real-time monitoring of quality. The signal-based approach is more suitable for offline quality diagnostics and can make use of more input information than the concurrent approach.

5.4 Chapter Summary

The study that was presented in this chapter was conducted to complete the knowledge about user perception of quality in heterogeneous networks, in this case when video telephony is used. Similarly to the work that was presented in the previous chapter, a set of user experiments was carried out, which allowed to explore the user perception of video telephony quality and of mobility management in wireless networks.

Based on the experience that was collected during the experiments that were conducted for speech telephony, the design of the video quality tests in this chapter was approached more systematically. However, once again the multitude of transmission aspects that was necessary to address in this study, in particular the number of test combinations due to time-varying quality, introduced a particular difficulty of the test design. Therefore, careful selection of test conditions was the main challenge of the test plan development, although not all conditions of interest could be covered.

In general, the user tests were designed to understand how an effect that occurs in both, audio and video channel of a video call, is perceived and how it affects the overall Quality of Experience of a video call. Therefore, the work presented in this chapter was based on three types of tests, spanning video-only and audiovisual viewing tests with emulation of video calls. A conversational test was not conducted in this study due to the disadvantage of this test methodology that was discovered in the previous chapter. This type of test turned out to be challenging to collect reliable results when time-variation of many transmission properties should be explored simultaneously.

The results of the conducted tests confirmed that proper selection of the wireless network, video codec, and the transmission bit rate is essential to provide high Quality of Experience. As a matter of fact, if high quality should be delivered, a stable network connection, high quality speech and video coding, and high encoding bit rate are required. Therefore, when networking circumstances were unimpaired, video calls that employ H.264 at 1536 kbit/s achieved the best quality ratings in the test, and high quality of the speech signal, which is of particular relevance when the video quality is high as well, contributed to the result. However, any change of this reference configuration degraded the Quality of Experience, in many cases making a significant difference, which is related to the fact that long exposure to low quality of a video call was rated lower than a short sample of the same transmission.

Moreover, it was also confirmed that the position, duration, and strength of quality degradation in a video call affects the quality perception. This implies that a robust video codec is required in wireless networks if an unimpaired connection cannot be guaranteed. If any, even short, link layer degradation is expected, the use of a robust video codec, such as MPEG-4 in this work, is recommended at no significant loss of the overall quality of a video call. Moreover, it has to be considered that although the encoding bit rate enables to maximize the Quality of Experience, the perceived quality improvement is not proportional, showing lower relative improvement of quality, the higher the encoding bit rate is. Therefore, a trade-off between the utilized resources and the provided quality should be considered. However, the situation is different when connection throughput is limited. In such a case, an efficient coding scheme, such as H.264, helps to significantly improve the Quality of Experience.

Any transmission changes during high quality streaming are not recommended, although it might be advantageous to switch the network when a connection is initially degraded. Therefore, when stable transmission in either HSDPA or WiFi can be guaranteed, switching between those networks during a video call in general should be avoided. In this context, also the quality of the available network connection that results from the load of a public access point is relevant to assure high Quality of Experience while roaming across heterogeneous networks. Otherwise, a degradation due to a network handover is experienced, which can be masked by a low encoding bit rate or robust video coding. However, if a handover can be foreseen, proactive scheduling of the entire transmission in a lower quality, but robust, network allows to maximize the Quality of Experience.

When the quality is adapted during an ongoing video call, switching between codecs or bit rates improves or degrades the quality according to the logical expectations. However, it has to be considered that a reduction of the video quality has a stronger impact on the user perception than a quality improvement. Therefore, if a drastic reduction of quality can be foreseen, proactive utilization of a low quality configuration reduces the service management effort at no significant costs of the perceived quality. Otherwise, the use of quality adaptation techniques during a video call is always advantageous, but the quality improvement is significant only when a certain boost of quality can be achieved, the adaptation is used early in a call, or when the process is applied to counterbalance the visual effect of packet loss. In the last case, replacing the video codec with a more robust encoding scheme is the only technique that always assures a noticeable quality improvement, but it has to be emphasised that if a mobile user is leaving the coverage area, a network handover is the only possibility to maintain the connection. Therefore, if a connection is degraded it is recommended to adapt the quality by changing the video codec first, and to switch the network when a certain level of degradation is exceeded. Moreover, it was also shown that the reduction of the video encoding bit rate is not an effective technique of quality adaptation when link-layer impairments occur [17]. Despite all of these facts, once again if connection degradations and the need of transmission adaptation can be foreseen, proactive use of a robust video codec or permanent streaming in a lower-quality but robust network enable to deliver higher Quality of Experience than any other of the dynamic adaptation methods that were considered above. Moreover, switching between narrowband and wideband speech during an ongoing video call should be avoided, and alternatively constant narrowband quality should be provided. However, if the transmission must be adapted, i.e. when the streaming bit rate should be reduced to minimum, it is recommended to change the video configuration first, before the bandwidth of the speech signal is switched. Low video quality reduces the sensitivity of video telephony users to degradations that appear in the audio channel [18].

Although the above results enabled to explore the quality perception of video calls in heterogeneous wireless networks extensively, a set of new research questions that are relevant for quality management in mobile networks arised. First of all the trade-off between audio and video quality when transmission is adapted should be further analysed, as an interaction in both directions was discovered. High speech quality amplified the results of video quality adaptation and vice versa. Second, the impact of the interval between the moments when the quality is improved and degraded in a call should be explored more carefully to provide detailed guidelines when quality improvements are perceptually advantageous, if consecutive quality degradations are expected.

Nevertheless, based on the analysis that was presented in this chapter, similarly to the conclusions of the previous chapter, two groups of mobility management recommendations can be derived. While the first group defines a direct response to dynamically changing transmission circumstances, the second group enables to proactively optimize the service quality when more information is available.

When the information in a dynamic networking environment is limited and the quality of a video call should be adopted due to changing networking circumstances, the following rules should be applied to improve the Quality of Experience:

- Switching between network technologies should be avoided as long as possible, excluding the case the connection is initially degraded by packet loss.
- If the transmission quality is degraded by packet loss up to 5 %, the video codec should be switched to a more robust encoding scheme.
- If the transmission quality is degraded by packet loss equal or higher than 5 %, a network handover should be scheduled.
- Quality improvements through switching to a higher quality codec or higher encoding bit rate should be avoided, when degradation of quality later in a call is expected.

When more information is available, which makes proactive quality management possible, the following guidelines are recommended:

- If unimpaired transmission of a video call is not possible, a rapid degradation of quality is expected, and a robust wide-range connection is available, the wide-range link should be utilized permanently.
- If any transmission impairments are expected, a robust video codec should be used from the beginning of a call, and the least degraded wireless connection should be selected.
- If connection throughput is limited, use of an efficient encoding scheme is recommended in any case.
- If a switch of the speech encoding bandwidth is expected, constant provisioning of narrowband speech is recommended.
- Otherwise, a high quality speech and video codec should be selected, and its encoding bit rate should be adjusted to maximize the Quality of Experience.

The work presented in this chapter also addressed the accuracy of quality prediction models that are applicable for assessment of video telephony quality, although none of them was strictly designed for this type of service in mobile networks, where time-varying quality is expected. Four types of models were addressed, including speech, video, audiovisual integration models, as well call quality models, from which the last type of models was successfully validated for applicability in video telephony systems in this work.

In this context, the accuracy of the video quality prediction models in heterogeneous wireless networks is primary defined by the purpose of a model, being J.341 the most accurate model when various content types are addressed, and G.1070 the most accurate model when video telephony is addressed. The J.144 model has not produced satisfying results in the context of this work, which is also not the purpose of the model. Moreover, although both of the top performing video models were developed to capture transmission related degradations, packet loss was the main reason of the estimation error in all of the conducted tests. It was also discovered that the G.1070 model must be extended to cover instability of networking circumstances not expressed as random packet loss, which is of particular relevance when the quality of high bit rate video should be assessed. Moreover, this model needs also to be improved to enable assessment of time-varying quality, as simplistic averaging of quality estimates is not an accurate method of quality prediction.

In turn, the speech quality models provided quality estimates that are expressed in a wider range on the MOS scale than the user judgements, which indicates that when the speech quality of a video call should be measured with regular models, the quality estimates must be transformed to correctly express the quality in the audiovisual context of a video call. Moreover, signal-based models once again turned out to predict the speech quality more accurately than the parametric E-model. However, the reasons of the estimation error were similar for all models, being packet loss and speech codec, in particular including a changeover between wideband and narrowband speech, the two main reasons of a quality estimation error. Moreover, it was discovered that in case of audiovisual transmission also the network, in particular a network handover, in relationship with high video encoding bit rate were causing speech signal interruptions, which was another reason of quality assessment error that was in particular critical in case of the parametric quality prediction.

Finally, the analyzed call quality models enabled to reliably assess the quality of emulated video calls, having quality overestimation as the only reason of quality estimation error. This problem was particularly observed when either the transmission quality was substantially degraded, which is related to the aforementioned fact that long exposure to bad quality is perceived worse than a short sample of the same quality degradation, or when the quality of a call was rapidly decreasing. Therefore, the models must be improved for this particular use case, which is a common problem of both, speech and video call quality estimation.

All in all, it was empirically shown that a sophisticated selection of quality prediction models enables to reliably predict the quality of a video call in a fully-instrumental way. However, if models that cannot handle certain types of degradations are used together, the error of quality prediction might be amplified. For example, although the audiovisual integration model provided quality estimates that are close to user opinion, when predicted quality scores were integrated, errors that affect audio and video quality estimates were projected to the predicted audiovisual quality. However, in the case of signal-based quality modeling, this error compensated for the overestimation that was introduced when the Weiss model was used, and so it improved the results.

References

1. ITU-T Rec. P.910 (1998) Subjective video quality assessment methods for multimedia applications. In: International telecommunication union, Geneve, Switzerland
2. ITU-T Rec. P.911 (1998) Subjective audiovisual quality assessment methods for multimedia applications. In: International telecommunication union, Geneve, Switzerland
3. ETSI 102 506 v. 1.2.1 (2006) Speech processing, transmission and quality aspects (STQ); Estimating speech quality per call. In: European telecommunications standard institute, Sophia Antipolis, France
4. Belmudez B, Lewcio B, Möller S (2012) Call quality prediction for audiovisual time-varying impairments using simulated conversational structures, Submitted to Acta Acustica United with Acustica
5. Khan A, Sun L, Ifeachor E (2009) Impact of video content on video quality for video over wireless networks. In: Proceedings of international conference on autonomic and autonomous systems (ICAS), Spain, pp 277–282
6. Wiegand T, Schwarz H, Joch A, Kossentini F, Sullivan GJ (2003) Rate-constrained coder control and comparison of video coding standards. IEEE Trans Circuits Syst Video Technol 13(7):688–703
7. Website of ffmpeg. http://www.ffmpeg.org/. Accessed June 07 2008
8. 3GPP TS 26.071 (1999) Mandatory speech CODEC speech processing functions; AMR speech Codec; General description. 3GPP
9. ITU-T Rec. G.722.2 (2003) Wideband coding of speech at around 16kbit/s using adaptive multi-rate wideband (AMR-WB). In: International telecommunication union, Geneve, Switzerland
10. Huszak A, Imre S (2010) Analysing GOP structure and packet loss effects on error propagation in MPEG-4 video streams. In: Proceedings of international symposium on communications, control, and signal processing (ISCCSP). Dallas, USA, pp 1–5

11. Schwarz H, Marpe D, Wiegand T (2007) Overview of the scalable video coding extension of the H. 264/AVC standard. IEEE Trans Circuits Syst Video Technol 17(9):1103–1120
12. VQEG MM-FR1. Final report from the video quality experts group on the validation of objective models of multimedia quality assessment, Phase I, 2008
13. Belmudez B (2012) Verification of the ITU-T Rec. G.1070 model for video telephony for interactive scenarios and extension to wideband speech. ITU-T Contr. COM 12–192, Geneve, Switzerland
14. Gros L, Chateau N (2001) Instantaneous and overall judgements for time-varying speech quality: assessments and relationships. Acta Acustica United Acustica 87(3):367–377
15. Weiss B, Möller S, Raake A, Berger J, Ullmann R (2009) Modeling call quality for time-varying transmission characteristics using simulated conversational structures. Acta Acustica United Acustica 95(6):1140–1151
16. Rosenbluth J (1998) Testing the quality of connections having time varying impairments. ITU-T Contr. D.64. Geneve, Switzerland
17. Wenger S (2003) H. 264/AVC over IP. IEEE Trans Circuits Syst Video Technol 13(7):645–656
18. Beerends J, De Caluwe F (2012) The influence of video quality on perceived audio quality and vice versa. J Audio Eng Soc 47(5)

Chapter 6
Conclusions

In this book, I have shown that networking research and quality engineering should be combined to assure high Quality of Experience in modern networks. An ideal communication system does not exist and many fundamental trade-offs must be considered when speech and video telephony services should be delivered to nomadic users. As a matter of fact, intelligent management of quality is required, and to be successfully in doing so, Quality of Experience in heterogeneous networks must be explored, and user perception of changing quality must be understood. These issues have been addressed in this book. The main results of the presented research and some directions of future work are summarized in this chapter.

6.1 Summary

This book was motived by the heterogeneity of transmission technologies that can be seamlessly used by nomadic users in modern communication systems. Through the transparent and ubiquitous access to speech and video telephony services, users can move through geographical areas that are covered by diverse wireless networks, while a connection to the service is preserved. To make the best of available resources, intelligent management of ongoing transmission is required and many fundamental trade-offs in provisioning the quality must be considered. Network coverage versus throughput and reliability of a connection is one key aspect, efficiency versus robustness of signal compression is another. In this context, the success of speech and video telephony offerings is tightly coupled with the Quality of Experience and with the awareness of user perception of changing quality during active sessions.

Therefore, to explore the Quality of Experience in heterogeneous wireless networks a research testbed was developed in the first stage of this work. The setup consists of a network infrastructure, of a speech and video telephony client, and of a control module. All these components enable to emulate transmission phenomena of

B. Lewcio, *Management of Speech and Video Telephony Quality in Heterogeneous Wireless Networks*, T-Labs Series in Telecommunication Services,
DOI: 10.1007/978-3-319-02102-7_6,

interest under controlled laboratory conditions in a passive as well as in an interactive test situation, and any effects that are perceived by the users can be captured.

The testbed enables to adjust a variety of transmission parameters in the test network and application, in particular enabling the control of those parameters at run-time. So, roaming between heterogeneous wireless technologies is supported, and networking degradations, for example due to packet loss or variable delay, can be emulated. Moreover, a multitude of speech and video codecs was integrated in the test client, and thus diverse speech and video coding configurations, in particular the trade-off between audio and video quality, can be evaluated. For this purpose, a technique to switch the speech and video codecs during an active call was developed, which is a core feature of the testbed. The transition between the codecs is seamless, active sessions are not interrupted, and signal decoding artifacts are reduced as far as possible. To achieve this goal a set of requirements that a seamless codec changeover has to fulfill was designed in scope of this work.

Despite the multitude of the supported features, the testbed was designed to reduce the complexity of performing a test. Therefore, a testbed controller is integrated in the setup, and an event-based concept of test definition was developed. In this way, any amount of data that is required to investigate the quality of arbitrary transmission scenarios can be generated, which includes both, stable and dynamic conditions, in which the streaming properties are changed several times during an ongoing call.

The above described testbed was used to carry out experiments that are necessary to understand user perception of speech and video telephony quality in heterogeneous wireless networks. Both of the studies were designed to understand how a single effect that might occur during a speech or a video call in these networks is perceived, and how it affects the overall Quality of Experience. In particular, user perception of quality changes due to switching between networks, codecs, and bit rates during an ongoing transmission is in focus of the analysis. For example, how a change of the speech codec is perceived, and what the effect of this single change is on the overall quality of a call. Therefore, various types of tests, spanning the listening-only and viewing-only situation, emulation of video calls, and real conversations, were conducted during this research.

The design of the test methodology in this work introduced certain research challenges. First of all, the multitude of transmission parameters that was in particular increased due to the combination of switching conditions imposed the main challenge of the test plan development. Therefore, although the test cases were carefully selected, not all of the conditions of interest could be explored. Moreover, user experiments that were based on real conversation made it difficult to collect reliable test results. Apart from the fact that users are less analytical in a conversational situation, the more dynamics are investigated in such a test, the greater is the risk of data unreliability. Therefore, a careful design of quality profiles that include a limited number of quality changes in a single call is recommended for future research.

The collected results revealed some similarities but also certain differences in quality perception when speech and video telephony is used in heterogeneous systems. As a matter of fact, proper selection of the wireless network, codec, and encoding bit rate is essential to provide high quality of a call. However, it should be considered that although the encoding bit rate enables to maximize the Quality of Experience, the quality improvement that is perceived by the users is not proportional to the utilized resources [1]. Moreover, long exposure to low quality is perceived worse than a short sample of the same quality [2]. Therefore, any discrepancy in the reference quality can have a significant effect in a long run.

When the quality of a call is impaired, the perceived effect is related to the position, duration, and strength of the degradation. However, if any connection related impairments are expected, robust signal codecs should be used permanently, at no significant costs of degraded quality, as compared to actively alternating between efficient and robust coding. In fact, when a connection is degraded, coding robustness is even more important than the encoded speech bandwidth.

When transmission is adapted during an ongoing call, it has to be considered that a reduction of quality is experienced quicker and stronger than a quality improvement [3]. The latter requires considerably more time to generate perceptual profits. Therefore, if a drastic reduction of quality can be foreseen, proactive utilization of a low-quality but robust configuration reduces the service management effort at no significant costs of the perceived quality. Otherwise, the use of quality adaptation techniques during a call is in most of the cases advantageous, but the quality improvement is substantial only when an essential boost of quality can be achieved, the adaptation is used early in a call, or when the process is applied to counterbalance the effect of packet loss. The only exception of this trend is observed when the speech codec is switched between narrowband and wideband quality. Such a switch itself degrades the perceived quality below the quality of pure narrowband in most of the cases, even if improvement of quality would be expected when the encoding bandwidth is switched to wideband. Therefore, a switch between wideband and narrowband quality is not recommended, and pure narrowband quality should be proactively provided if such a negatively perceived event is expected. Only when it is switched early in a call, a switch to wideband can help to significantly improve the user experience.

A particular diversity of the perceived effects was discovered in this study when the impact of the network technology and network handover on user perception was addressed. A handover to a loss free network turned out to be a powerful tool for adaptation of quality only when speech telephony was used. Despite this fact, a handover during a video call improved the Quality of Experience only if the initial quality degradation was strong. Otherwise the process degraded the Quality of Experience proportionally to the transmission bit rate. Therefore, when degradation of quality is expected, and a robust wide-coverage network is available, this link should be used proactively. Otherwise, if a connection is degraded, it is recommended to improve the coding robustness first, and to switch the network when a certain level of connection degradation according to the type of service is exceeded. The phase when a robust codec is used can also help to overcome the activation of a new network interface before a handover. Finally, it was also proven that the reduction of the video encoding bit rate is not an effective technique of quality adaptation when link-layer impairments occur [4]; examples of algorithms that schematically apply the derived guidelines are depicted in Fig. 6.1.

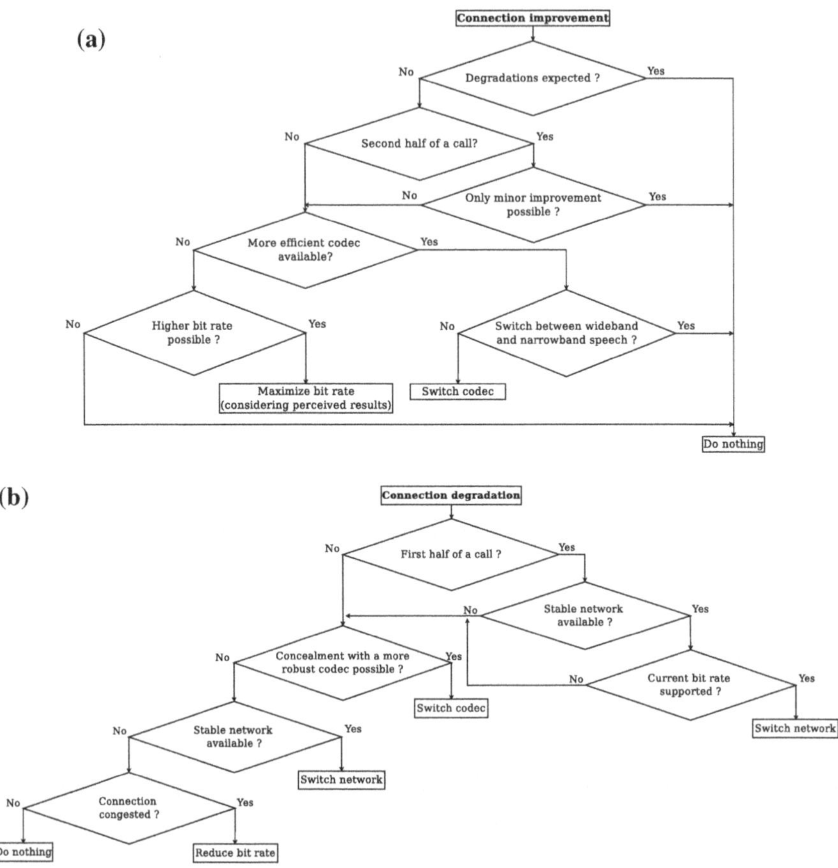

Fig. 6.1 Examples of management algorithms that implement the derived guidelines, when the quality of a network connection is changing. **a** Connection improvement, **b** Connection degradation

As a result, the knowledge of user perception in heterogeneous wireless networks does not only help to improve the Quality of Experience. It also enables the service operators to optimize the invested resources, and to reduce the effort of service management in certain situations. In this way, adding perception-awareness in the management protocols may also improve the scalability of a mobile telecommunication system. Accordingly, the rules derived in this thesis can be classified in two groups, active and proactive recommendations of service management. While the first set of rules was designed to maximize the Quality of Experience when the quality has to be adapted dynamically to changing transmission circumstances, the second set of rules enables to proactively optimize the service quality and invested resources when more contextual information about the transmission is available.

The results of the conducted user tests were also used to validate how accurately can the quality in heterogeneous networks be predicted with existing quality models.

In the domain of speech, only the signal-based models, POLQA and WB-PESQ provided quality estimates that were close to user opinion. However, both of them showed a weakness in the assessment of conditions that were affected by a changeover between narrowband and wideband speech codecs, or by strong and instable packet loss degradations. In turn, simplistic averaging of quality estimates that were provided by the E-model did not enable to reliably predict the quality. Therefore, the model was extended and a codec switching impairment factor was introduced. This modification enables to parametrically predict the quality of speech samples that are affected by a switch between different speech encoding bandwidths, which was in particular trained for the case of a changeover between wideband and narrowband.

Moreover, when speech quality models were used outside of their typical use-case, two additional observations were made. First, when applied in an error-free multi-band system, where narrowband, wideband, and super-wideband speech is simultaneously supported, and the transmission is never impaired by packet loss, the quality prediction models failed to benchmark pure coding quality. Apparently, strict training of super wideband models for prediction of transmission impairments limits their use-case to such a situation only. Second, when speech quality models were applied to assess the quality of a video call, the speech quality was systematically miss estimated. This effect is a result of the interaction between the speech and video quality in a video call. Because user judgements of speech quality in such a situation are compressed in a narrower range of the scale, additional mapping of the predicted values is required.

In turn, the accuracy of the video quality models in heterogeneous wireless networks is primarily defined by the purpose of a model, being J.341 the most accurate model when various content types are addressed, and G.1070 the most accurate model when video telephony is in focus. The J.144 model has not produced satisfying results for the applications in this work, which is also not the purpose of that model. Moreover, although both of the top performing video models were developed to capture transmission related degradations, packet loss was the main reason of the estimation error. In addition, the G.1070 model did not allow to estimate quality degradations due to instable networking circumstances that are not expressed as a random packet loss. Therefore, the model must be also extended to enable assessment of time-varying quality, because simplistic averaging of quality estimates was not an accurate method of quality prediction in this case.

Finally, it has been empirically proven that the existing call quality models enable to reliably assess the quality of both, speech and video calls, in heterogeneous wireless networks. Although all of the tested methods provided quality estimates close to the user perception, models that included additional weighting of bad transmission periods during a call were more accurate. Nevertheless, if the quality of a call was substantially degraded or it was rapidly decreasing, the accuracy of the models was reduced. This might be related to the aforementioned fact that long exposure to bad quality is perceived worse than a short sample of the same degradation. Both of the aspects must be further improved, which is of particular relevance in heterogeneous wireless networks where sudden quality degradations are expected.

6.2 Future Work

This book has raised a set of new issues that have yet to be addressed to improve the Quality of Experience in modern telecommunication systems. The most interesting directions of future work are discussed below.

First of all, it is necessary to understand the interaction between improvements and degradations of quality that can happen within a single call. In this study it was generally shown, that hopping between different quality levels is not recommended. However, the impact of the interval between the moments when the quality is improved and degraded in a call should be explored further to provide detailed guidelines when quality improvements are perceptually advantageous, if consecutive quality degradations are expected.

Moreover, it has to be analysed whether a concealment of a direct switch between two different quality levels helps to improve the Quality of Experience. For example, the duration of a quality switch can be extended and the transition can be performed in discrete steps that correspond to so called "just noticeable differences" that are below the threshold of a quality change that can be directly detected by a user.

Moreover, additional parameters that are related to quality perception and were not addressed in this work must be explored to further optimize the quality management. For example, user perception of quality adaptation in terms of video resolution and frame rate are yet further characteristics that should be analysed, in particular when the transmission is degraded.

Moreover, the interaction between speech and video quality when the transmission is adapted should be understood. In this work, the problem was discovered that high speech quality amplified the perceptual effects of video quality adaptation and vice versa. Therefore, more research is required to provide transparent recommendations for audiovisual quality adaptation during a video call.

Finally, the last but thus the most interesting direction of future research is to develop a prototype of a mobility management system that combines the knowledge that is presented in this book with the concept of cross-layer service adaptation, and with the network-awareness as proposed in IEEE 802.21. Such a system would not only improve the Quality of Experience of nomadic users, but it would also contribute to optimized use of resources in the intelligent networks of the future.

References

1. Belmudez B, Möller S (2011) An approach for modeling the effects of video resolution and size on the perceived visual quality. In: Proceedings of IEEE workshop on multimedia quality of experience (MQoE), Dana Point, USA, pp 464–469
2. Rosenbluth J (1998) Testing the quality of connections having time varying impairments. ITU-T Contr D.64, Geneve, Switzerland
3. Gros L, Chateau N (2001) Instantaneous and overall judgements for time-varying speech quality: assessments and relationships. Acta Acustica United with Acustica 87(3):367–377
4. Wenger S (2003) H. 264/AVC over IP. IEEE Trans Circuits Syst Video Technol 13(7):645–656

Appendix A

A.1 Data from Conducted Speech Quality Tests

Table A.1 Statistics of all the conducted listening-only speech tests

Name	Test S1a	Test S1b	Test S2a	Test S2b	Test S3a	Test S3b
No. of invited participants	13	24	14	17	20	20
No. of rejected participants	2	0	4	0	3	2
Cronbach's alpha	0.964	0.975	0.958	0.811	0.783	0.922

Table A.2 Description of call quality profiles in Test S1a, which includes the profile name, as well as the network, codec, and packet loss configuration, and the number of sample in the middle of which a parameter was switched

	Name	Network	Codec	Loss	Sample
1	Stable NB	HSDPA	G.711	0	-
2	Stable WB	WiFi	G.722.2	0	-
3	NB to WB beg.	HSDPA to WiFi	G.711 to G.722.2	0	2nd
4	NB to WB end	HSDPA to WiFi	G.711 to G.722.2	0	4th
5	WB to NB beg.	WiFi to HSDPA	G.722.2 to G.711	0	2nd
6	WB to NB end	WiFi to HSDPA	G.722.2 to G.711	0	4th
7	NB (loss)	HSDPA	G.711	0,10,20,10,10	-
8	WB (loss)	WiFi	G.722.2	0,10,20,10,10	-
9	NB (loss) to WB mid.	HSDPA to WiFi	G.711 to G.722.2	0,10,20,0,0	3rd
10	WB (loss) to NB mid.	WiFi to HSDPA	G.722.2 to G.711	0,10,20,0,0	3rd

B. Lewcio, *Management of Speech and Video Telephony Quality in Heterogeneous Wireless Networks*, T-Labs Series in Telecommunication Services,
DOI: 10.1007/978-3-319-02102-7, © Springer International Publishing Switzerland 2014

Table A.3 Quality scores in Test S1a

	Name	MOS_a	CI_{up}	CI_{low}	ETSI	Weiss	Rosenbluth	Mixed	Parametric
1	Stable NB	3.11	3.30	2.93	2.81	2.81	2.81	2.89	3.67
2	Stable WB	3.57	3.78	3.36	4.25	4.25	4.25	3.91	4.42
3	NB to WB beg.	3.25	3.44	3.06	3.68	3.34	3.41	3.31	3.85
4	NB to WB end	2.82	3.04	2.60	3.15	3.00	2.86	3.07	3.67
5	WB to NB beg.	2.91	3.09	2.73	2.97	2.75	2.83	2.85	3.60
6	WB to NB end	3.05	3.24	2.85	3.49	3.10	3.05	3.09	3.78
7	NB (loss)	2.27	2.55	2.00	1.97	1.79	1.85	2.12	2.80
8	WB (loss)	2.07	2.28	1.86	2.23	1.91	1.90	2.21	2.49
9	NB (loss) to WB mid.	2.91	3.13	2.69	3.12	2.75	2.51	2.78	3.42
10	WB (loss) to NB mid.	2.66	2.89	2.43	2.65	2.27	2.34	2.46	2.85

Table A.4 Description of conditions in Test S1b, including the relative position if the codec was switched before or after the network

	$Network_1$	$Network_2$	$Codec_1$	$Codec_2$	$Loss_1$	$Loss_2$	Position
1	WiFi	WiFi	G.722.2	G.722.2	0.00	0.00	
2	WiFi	WiFi	G.722.2	G.722.2	10.00	10.00	
3	WiFi	WiFi	G.722.2	G.722.2	20.00	20.00	
4	HSDPA	HSDPA	G.711	G.711	0.00	0.00	
5	HSDPA	HSDPA	G.711	G.711	10.00	10.00	
6	HSDPA	HSDPA	G.711	G.711	20.00	20.00	
7	HSDPA	HSDPA	G.711	G.722.2	0.00	0.00	
8	HSDPA	HSDPA	G.722.2	G.711	0.00	0.00	
9	WiFi	HSDPA	G.722.2	G.722.2	0.00	0.00	
10	Flarion	HSDPA	G.722.2	G.722.2	0.00	0.00	
11	HSDPA	WiFi	G.711	G.711	0.00	0.00	
12	HSDPA	Flarion	G.711	G.711	0.00	0.00	
13	WiFi	HSDPA	G.722.2	G.711	0.00	0.00	after
14	WiFi	HSDPA	G.722.2	G.711	0.00	0.00	before
15	Flarion	HSDPA	G.722.2	G.711	0.00	0.00	after
16	Flarion	HSDPA	G.722.2	G.711	0.00	0.00	before
17	HSDPA	WiFi	G.711	G.722.2	0.00	0.00	after
18	HSDPA	WiFi	G.711	G.722.2	0.00	0.00	before
19	HSDPA	Flarion	G.711	G.722.2	0.00	0.00	after
20	HSDPA	Flarion	G.711	G.722.2	0.00	0.00	before
21	HSDPA	WiFi	G.711	G.711	10.00	0.00	
22	WiFi	HSDPA	G.722.2	G.722.2	10.00	0.00	
23	Flarion	HSDPA	G.722.2	G.722.2	10.00	0.00	
24	HSDPA	Flarion	G.711	G.711	10.00	0.00	
25	WiFi	HSDPA	G.722.2	G.711	20.00	0.00	before
26	HSDPA	WiFi	G.711	G.722.2	20.00	0.00	before

Table A.5 Quality scores in Test S1b

	MOS_a	CI_{up}	CI_{low}	RMOS	$RMOS_{int}$	PESQ	POLQA	E-model	R	$E\text{-model}_{enh}$	R_{enh}
1	4.25	4.41	4.09	111.84	121.15	3.74	3.91	4.42	121.00	4.42	121.00
2	1.95	2.11	1.78	48.50	72.04	1.65	2.27	2.42	60.77	2.42	60.77
3	1.20	1.29	1.11	23.31	52.51	1.23	1.67	2.00	49.92	2.00	49.92
4	2.81	3.01	2.62	70.32	88.96	2.68	2.89	3.67	92.35	3.67	92.35
5	1.90	2.05	1.74	47.07	70.94	1.77	2.15	2.98	74.39	2.73	68.15
6	1.39	1.50	1.27	31.15	58.59	1.60	1.86	2.63	65.76	2.63	65.76
7	2.24	2.39	2.09	56.14	77.96	1.47	2.38	4.05	106.51	3.27	81.58
8	2.20	2.36	2.04	55.08	77.14	1.70	2.24	4.06	107.10	3.29	82.16
9	4.04	4.22	3.87	103.81	114.92	3.49	3.81	4.42	121.00	4.42	121.00
10	3.85	4.04	3.67	97.73	110.21	3.44	3.85	4.42	121.00	4.42	121.00
11	3.17	3.34	3.00	79.07	95.74	2.94	3.06	3.87	99.94	3.49	87.47
12	2.97	3.15	2.79	74.15	91.93	2.66	2.77	3.69	93.04	3.69	93.04
13	2.23	2.39	2.07	55.87	77.76	1.72	2.70	4.06	107.10	3.29	82.16
14	2.25	2.37	2.13	56.40	78.17	2.01	2.50	4.06	107.10	3.29	82.16
15	2.38	2.55	2.20	59.55	80.61	1.84	2.82	4.06	107.10	3.29	82.16
16	2.44	2.62	2.25	61.11	81.82	1.81	2.83	4.06	107.10	3.29	82.16
17	2.45	2.61	2.29	61.36	82.02	1.80	2.65	4.03	105.80	3.24	80.87
18	2.51	2.62	2.40	62.91	83.21	1.95	2.79	4.06	107.10	3.29	82.16
19	2.51	2.67	2.35	62.91	83.21	1.84	2.46	4.03	106.24	3.25	81.30
20	2.35	2.51	2.20	59.03	80.21	1.87	2.83	4.06	107.10	3.29	82.16
21	2.14	2.30	1.98	53.47	75.89	1.93	1.98	3.33	83.65	2.59	64.95
22	2.20	2.37	2.02	55.08	77.14	2.02	2.45	3.47	92.06	3.66	92.06
23	2.21	2.37	2.05	55.34	77.35	1.83	2.50	3.48	92.25	3.67	92.25
24	2.33	2.51	2.16	58.51	79.80	2.04	2.24	3.37	84.40	3.37	84.40
25	1.35	1.46	1.25	29.97	57.68	1.24	1.76	2.84	71.27	1.87	46.33
26	1.78	1.92	1.64	43.85	68.44	1.42	2.02	3.59	95.00	2.80	70.06

Table A.6 Description of call quality profiles in Test S2a, which includes the profile name, as well as the network, codec, and packet loss configuration, and the number of sample in the middle of which a parameter was switched

	Name	Network	Codec	Loss	Sample
1	Stable NB	HSDPA	G.711	0	-
2	Stable WB	WiFi	G.722.2	0	-
3	NB to WB beg.	HSDPA to WiFi	G.711 to G.722.2	0	2
4	NB to WB mid.	HSDPA to WiFi	G.711 to G.722.2	0	3
5	WB (3 % loss) to NB mid.	WiFi to HSDPA	G.722.2 to G.711	0,3,3,0,0	3
6	WB (5 % loss) to NB mid.	WiFi to HSDPA	G.722.2 to G.711	0,5,5,0,0	3
7	WB (10 % loss) to NB mid.	WiFi to HSDPA	G.722.2 to G.711	0,10,10,0,0	3
8	WB to NB to WB	WiFi and HSDPA	G.722.2 and G.711	0	2,4
9	WB (5 % loss mid.)	WiFi	G.722.2	0, 0 to 5, 5, 5 to 0, 0	2,4
10	NB to WB to NB to WB	WiFi and HSDPA	G.722.2 and G.711	0	2,3,4

Table A.7 Quality scores in Test S2a. This table includes user judgements only, because this experiment was not used for validation of the quality prediction models

	Name	MOS_a	CI_{up}	CI_{low}
1	Stable NB	3.30	3.59	3.01
2	Stable WB	4.00	4.28	3.72
3	NB to WB beg.	3.65	3.90	3.40
4	NB to WB mid.	3.38	3.63	3.12
5	WB (3 % loss) to NB mid.	2.85	3.05	2.65
6	WB (5 % loss) to NB mid.	2.52	2.75	2.30
7	WB (10 % loss) to NB mid.	2.33	2.58	2.07
8	WB to NB to WB	2.92	3.19	2.66
9	WB (5 % loss mid.)	3.02	3.39	2.66
10	NB to WB to NB to WB	2.85	3.14	2.56

Table A.8 Description and quality scores of conditions in Test 2b. This table includes all relevant data of the experiment, including MOS scores, because Test 2b was not used for validation of the quality prediction models

	$Network_1$	$Network_2$	$Codec_1$	$Codec_2$	$Loss_1$	$Loss_2$	MOS_a	CI_{up}	CI_{low}
1	WiFi	WiFi	G.722.2	G.722.2	0.00	0.00	4.15	4.26	4.04
2	WiFi	WiFi	G.722.2	G.722.2	3.00	3.00	3.65	3.93	3.36
3	WiFi	WiFi	G.722.2	G.722.2	5.00	5.00	2.63	2.85	2.42
4	WiFi	WiFi	G.722.2	G.722.2	10.00	10.00	2.15	2.43	1.86
5	WiFi	WiFi	G.722.2	G.722.2	0.00	5.00	3.09	3.38	2.80
6	WiFi	WiFi	G.722.2	G.722.2	5.00	0.00	3.29	3.63	2.96
7	HSDPA	HSDPA	G.711	G.711	0.00	0.00	3.43	3.53	3.33
8	WiFi	HSDPA	G.722.2	G.711	0.00	0.00	2.75	2.94	2.56
9	WiFi	HSDPA	G.722.2	G.711	3.00	0.00	2.47	2.77	2.17
10	WiFi	HSDPA	G.722.2	G.711	5.00	0.00	2.76	2.99	2.54
11	WiFi	HSDPA	G.722.2	G.711	10.00	0.00	2.26	2.46	2.07
12	HSDPA	WiFi	G.711	G.722.2	0.00	0.00	2.96	3.09	2.82

Table A.9 Description of call quality profiles in Test 3a, which includes the profile name, as well as the network, codec, and packet loss configuration, and the number of sample in the middle of which a parameter was switched

	Name	Network	Codec	Loss	Sample
1	Stable NB	WiFi	G.711	0.00	–
2	Stable WB	WiFi	G.722	0.00	–
3	Stable SWB	WiFi	G.721.C	0.00	–
4	WB to SWB beg.	WiFi	G.722 to G.722.1C	0.00	1st
5	WB to SWB mid.	WiFi	G.722 to G.722.1C	0.00	3rd
6	WB to SWB end	WiFi	G.722 to G.722.1C	0.00	5th
7	SWB to WB beg.	WiFi	G.722.1C to G.722	0.00	1st
8	SWB to WB mid.	WiFi	G.722.1C to G.722	0.00	3rd
9	SWB to WB end	WiFi	G.722.1C to G.722	0.00	5th
10	NB to WB mid.	WiFi	G.711 to G.722	0.00	3rd
11	WB to NB mid.	WiFi	G.722 to G.711	0.00	3rd
12	NB to WB to SWB	WiFi	G.711 to G.722 to G.722.1C	0.00	2nd, 4th
13	SWB to WB to NB	WiFi	G.722.1C to G.722 to G.711	0.00	2nd, 4th
14	NB to SWB mid.	WiFi	G.711 to G.722.1C	0.00	3rd
15	SWB to NB mid.	WiFi	G.722.1C to G.711	0.00	3rd

Table A.10 Quality scores in Test 3a

	Name	MOS_a	CI_{up}	CI_{low}	ETSI	Weiss	Rosenbluth	Mixed	Parametric
1	Stable NB	2.37	2.57	2.18	2.53	2.53	2.53	3.35	3.27
2	Stable WB	3.41	3.66	3.16	3.69	3.69	3.69	3.94	4.17
3	Stable SWB	4.06	4.29	3.83	4.22	4.22	4.22	4.21	4.47
4	WB to SWB beg.	3.96	4.13	3.79	4.16	4.07	4.15	4.06	4.27
5	WB to SWB mid.	3.81	3.99	3.63	3.97	3.89	3.92	4.00	4.20
6	WB to SWB end	3.60	3.80	3.40	3.74	3.73	3.73	3.90	4.08
7	SWB to WB beg.	3.81	3.99	3.63	3.71	3.70	3.71	3.87	4.10
8	SWB to WB mid.	3.54	3.73	3.36	3.90	3.83	3.88	3.94	4.17
9	SWB to WB end	3.81	3.98	3.64	4.10	3.99	4.10	3.99	4.22
10	NB to WB mid.	2.85	3.07	2.63	3.06	2.92	2.92	3.39	3.39
11	WB to NB mid.	2.44	2.63	2.25	2.93	2.76	2.74	3.53	3.31
12	NB to WB to SWB	3.29	3.53	3.05	3.45	3.19	3.25	3.51	3.54
13	SWB to WB to NB	2.53	2.73	2.32	3.25	2.98	2.95	3.61	3.43
14	NB to SWB mid.	3.04	3.24	2.85	3.33	3.12	3.02	3.45	3.38
15	SWB to NB mid.	2.29	2.48	2.11	3.14	2.94	2.83	3.60	3.27

Table A.11 Description of conditions in Test S3b, which includes the network, codec, and packet loss configuration before and after the switch. If the corresponding values are equal, no switch was performed, and in case of switching of two parameters, codec was always switched after the network handover

	$Network_1$	$Network_2$	$Codec_1$	$Codec_2$	$Loss_1$	$Loss_2$
1	WiFi	WiFi	G.711	G.711	0.00	0.00
2	WiFi	WiFi	G.722	G.722	0.00	0.00
3	WiFi	WiFi	G.722.1C	G.722.1C	0.00	0.00
4	None	None	Uncompressed	Uncompressed	0.00	0.00
5	WiFi	WiFi	G.722	G.722.1C	0.00	0.00
6	WiFi	WiFi	G.722.1C	G.722	0.00	0.00
7	WiFi	WiFi	G.711	G.722	0.00	0.00
8	WiFi	WiFi	G.722	G.711	0.00	0.00
9	WiFi	WiFi	G.711	G.722.1C	0.00	0.00
10	WiFi	WiFi	G.722.1C	G.711	0.00	0.00

Table A.12 Quality scores in Test S3b

	MOS_a	CI_{up}	CI_{low}	RMOS	POLQA	E-model	$E\text{-model}_{enh}$
1	2.53	2.74	2.31	93.30	3.35	3.27	3.27
2	3.69	3.87	3.52	137.03	3.94	4.17	4.17
3	4.22	4.42	4.02	162.98	4.21	4.47	4.47
4	4.33	4.53	4.12	170.09	4.57	4.50	4.50
5	3.86	4.03	3.69	144.27	3.87	4.32	4.00
6	3.76	3.93	3.60	139.98	3.79	4.32	4.00
7	2.51	2.70	2.32	92.79	2.98	3.72	2.84
8	2.42	2.59	2.24	89.25	3.63	3.72	2.84
9	2.67	2.85	2.49	98.32	2.90	3.87	2.58
10	2.54	2.74	2.34	93.80	3.67	3.87	2.58

Table A.13 Description of quality profiles in Test S4, which includes the profile name, as well as the network, codec, and packet loss configuration, and the duration of particular segments, if a parameter was switched

	Name	Codec	Loss	Switch
1	WiFi	G.722	0	-
2	HSDPA	G.711	0	-
3	WiFi[45s] to WiFi(3 %)[30 s] to HSDPA	G.722 to G.722 to G.711	0 to 3 to 0	45 s, 30 s
4	WiFi[45s] to WiFi(5 %)[30 s] to HSDPA	G.722 to G.722 to G.711	0 to 5 to 0	45 s, 30 s
5	WiFi[45s] to WiFi(10 %)[30 s] to HSDPA	G.722 to G.722 to G.711	0 to 10 to 0	45 s, 30 s
6	WiFi[75s] to WiFi(5 %)[30 s] to WiFi	G.722	0 to 5 to 0	75 s, 30 s
7	WiFi[75s] to WiFi(5 %)	G.722	0 to 5	75 s, 30 s
8	WiFi[75s] to WiFi(10 %)	G.722	0 to 10	75 s, 30 s
9	WiFi[45s] to HSDPA[60 s] to WiFi	G722 to G.711 to G.722	0	45 s, 60 s
10	WiFi[10s] to HSDPA[100 s] to WiFi	G722 to G.711 to G.722	0	10 s, 100 s
11	WiFi[10s] to HSDPA[25 s] x 4 to WiFi	G.722 and G.711	0	(10 s, 25 s)x4
12	WiFi[10s] to HSDPA[25 s] x 4 to WiFi (G.722 only)	G.722	0	(10 s, 25 s)x4
13	WiFi[10s] to HSDPA[100 s] to WiFi (G.722 only)	G.722	0	10 s, 100 s
14	WiFi to HSDPA non-stop (G.722 only)	G.722	0	non-stop

Table A.14 Quality scores that were collected in Test S4 on the Correspondent and Mobile Node, including the number of ratings per profile

	Mobile			Correspondent			
	MOS_a	CI_{low}	CI_{up}	MOS_a	CI_{low}	CI_{up}	Count
1	4.00	3.45	4.55	3.85	3.30	4.39	13
2	2.85	2.30	3.39	3.31	2.59	4.02	13
3	2.38	1.92	2.85	2.54	2.07	3.01	13
4	2.69	2.24	3.15	2.23	1.62	2.84	13
5	2.67	2.25	3.08	2.25	1.77	2.73	12
6	4.20	3.64	4.76	4.10	3.39	4.81	10
7	2.55	1.79	3.30	4.18	3.68	4.69	11
8	2.92	2.34	3.49	3.83	3.13	4.54	12
9	2.91	2.21	3.61	2.18	1.68	2.69	11
10	3.00	2.33	3.67	2.60	2.10	3.10	10
11	2.67	1.40	3.94	1.83	1.04	2.62	6
12	3.86	3.03	4.69	3.29	2.83	3.74	7
13	4.12	3.43	4.82	3.38	2.49	4.26	8
14	4.08	3.56	4.60	3.31	2.51	4.10	13

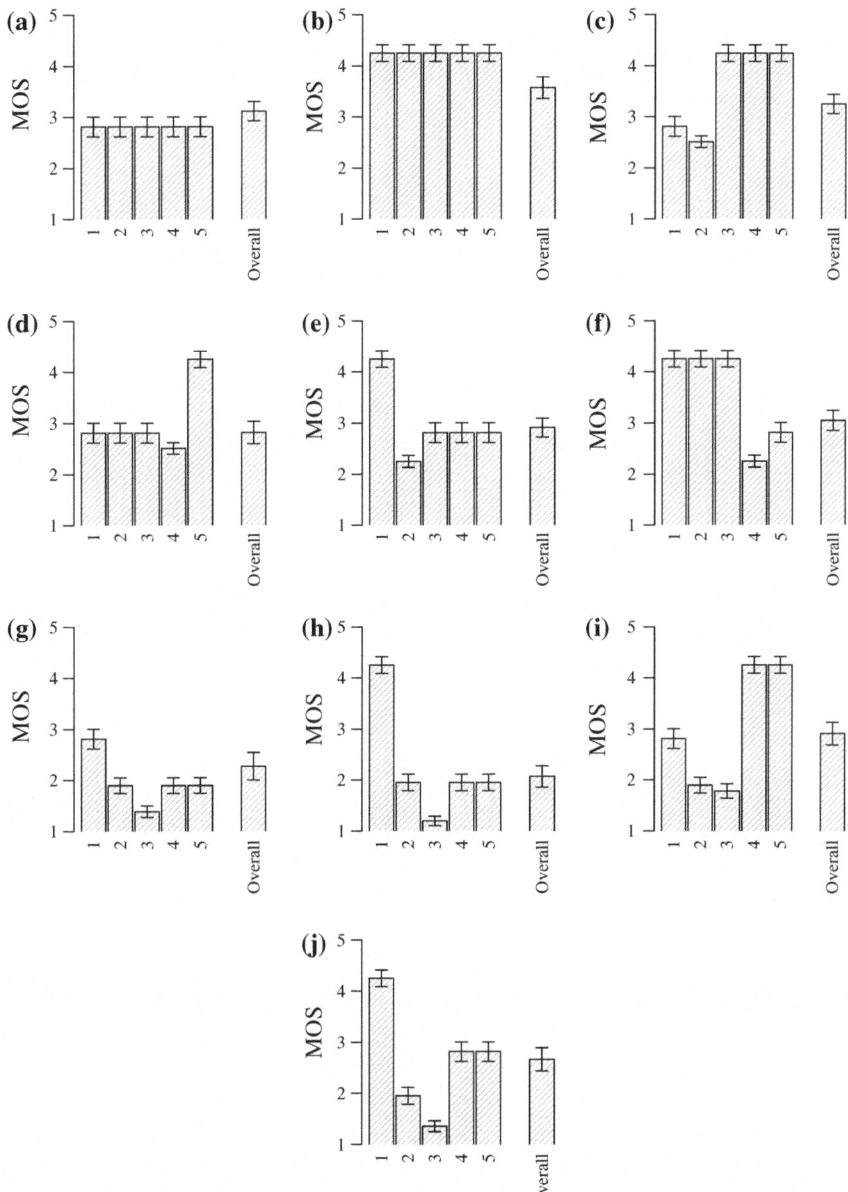

Fig. A.1 Call quality profiles in Test S1a. MOS_a and 95 % CI. **a** Stable NB. **b** Stable WB. **c** NB to WB beg. **d** NB to WB end. **e** WB to NB beg. **f** WB to NB end. **g** NB (loss). **h** WB (loss). **i** NB (loss) to WB mid. **j** WB (loss) to NB mid

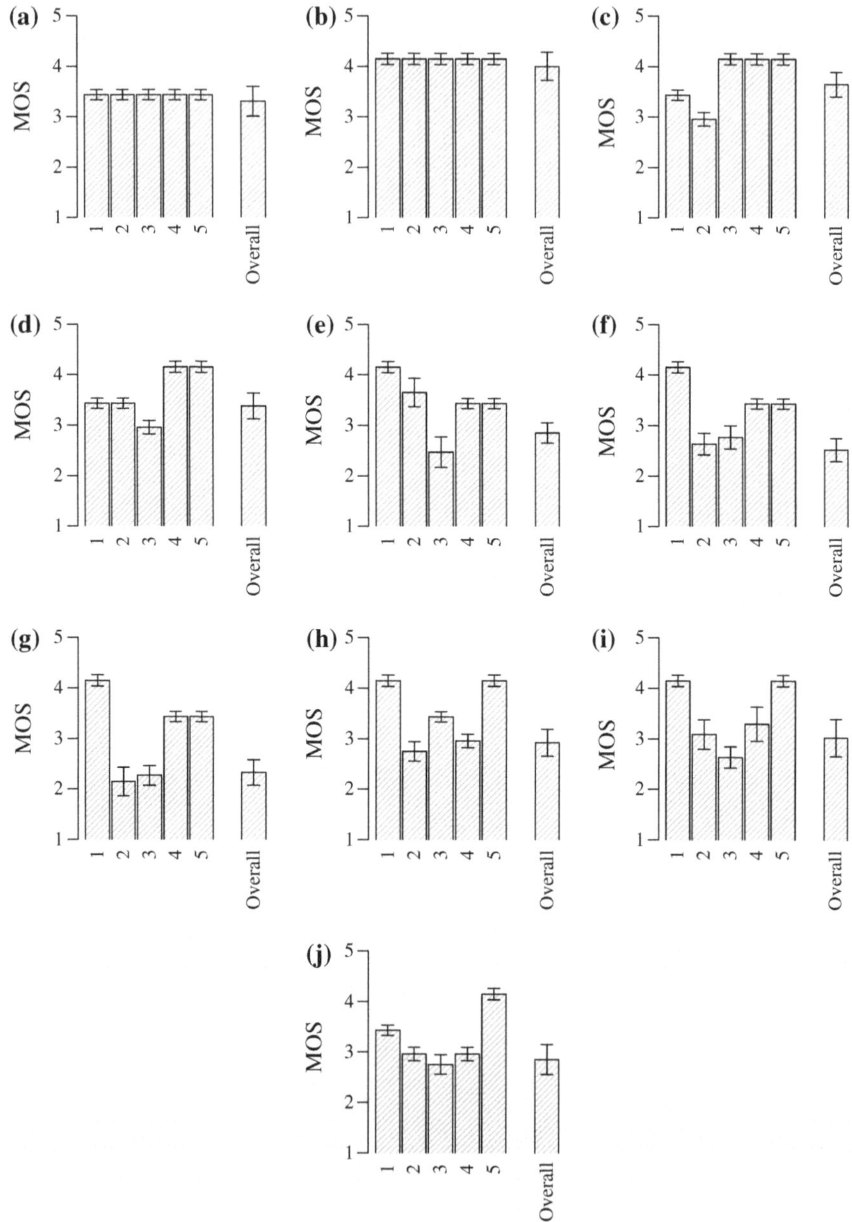

Fig. A.2 Call quality profiles in Test S2a. MOS_a and 95 % CI. **a** Stable NB. **b** Stable WB. **c** NB to WB beg. **d** NB to WB mid. **e** WB (3 % loss) to NB mid. **f** WB (5 % loss) to NB mid. **g** WB (10 % loss) to NB mid. **h** WB to NB to WB. **i** WB (5 % loss mid.). **j** NB to WB to NB to WB

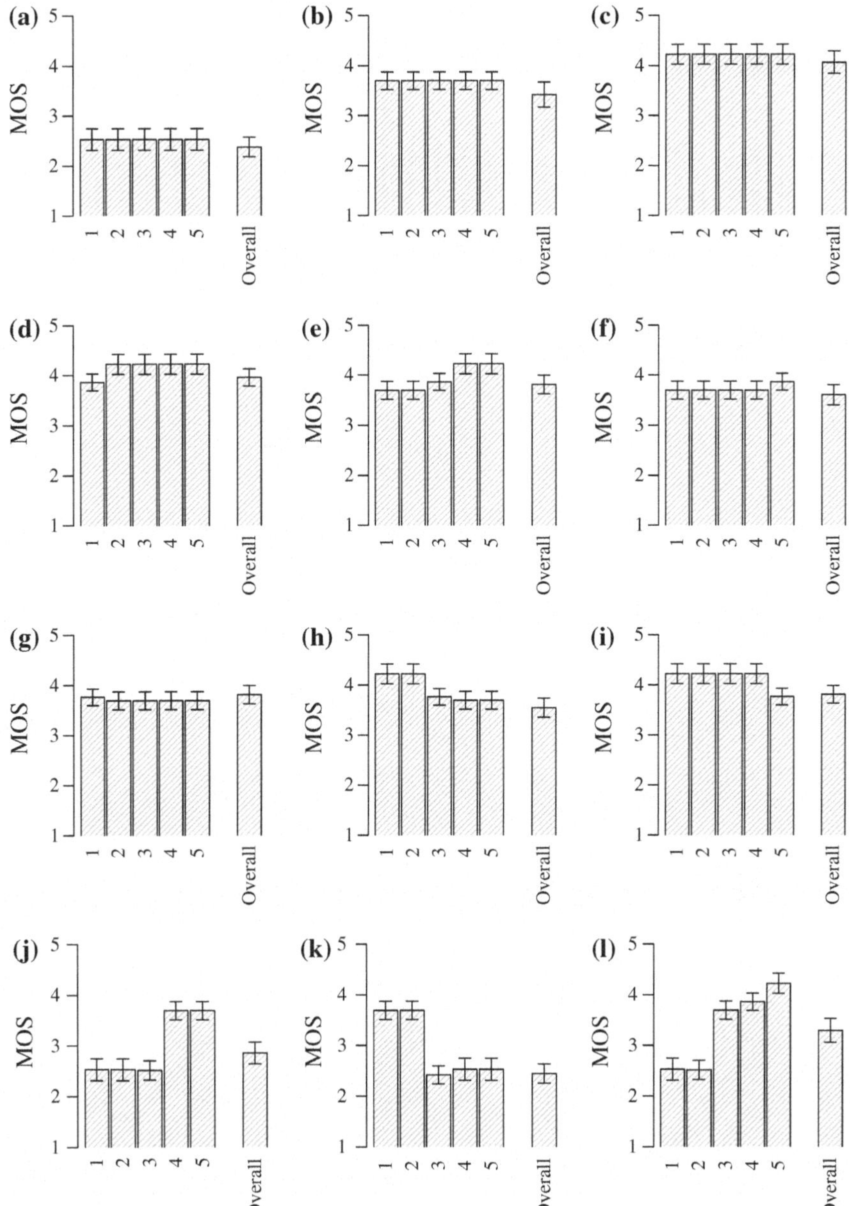

Fig. A.3 Call quality profiles of Test S3a. MOS_a and 95 % CI. **a** Stable NB. **b** Stable WB. **c** Stable SWB. **d** WB to SWB beg. **e** WB to SWB mid. **f** WB to SWB end. **g** SWB to WB beg. **h** SWB to WB mid. **i** SWB to WB end. **j** NB to WB mid. **k** WB to NB mid. **l** NB to WB to SWB. **m** SWB to WB to NB. **n** NB to SWB mid. **o** SWB to NB mid

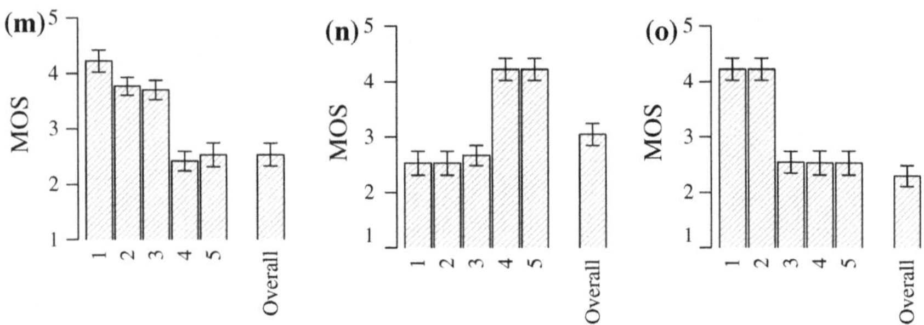

Fig. A.3 Continued

A.2 Data from Conducted Video Quality Tests

Table A.15 Statistics of all the conducted video tests

Name	Test V1	Test V2	Test V3
No. of invited participants	21	20	20
No. of rejected participants	0	1	1
Cronbach's alpha	0.971	0.931	0.924

Table A.16 Codec coefficients used for calculation in the video model from ITU-T Rec. G.1070

Parameter	H.264	MPEG-4
v1	6.9522	1.7079
v2	0.016	0.0104
v3	3.4109	5.3117
v4	105.6922	936.9469
v5	0.985	0.881
v6	1.1993	2.7072
v7	0.00042695	−0.00097085
v8	0.1141	0.8327
v9	388.0784	188.4333
v10	0.8032	5.1957
v11	0.9854	208.7758
v12	11.2699	272.6761

Table A.17 Description of conditions in Test V1, which includes network, codec, video encoding bit rate, and and packet loss configuration before and after the switch. If the corresponding values are equal, no switch was performed

	$Network_1$	$Network_2$	$VCodec_1$	$VCodec_2$	$VBitrate_1$	$VBitrate_2$	$Loss_1$	$Loss_2$
1	WiFi	HSDPA	MPEG-4	MPEG-4	256.00	256.00	0.00	0.00
2	WiFi	HSDPA	MPEG-4	MPEG-4	768.00	768.00	0.00	0.00
3	WiFi	HSDPA	MPEG-4	MPEG-4	1536.00	1536.00	0.00	0.00
4	WiFi	HSDPA	H.264	H.264	256.00	256.00	0.00	0.00
5	WiFi	HSDPA	H.264	H.264	768.00	768.00	0.00	0.00
6	WiFi	HSDPA	H.264	H.264	1536.00	1536.00	0.00	0.00
7	HSDPA	WiFi	MPEG-4	MPEG-4	256.00	256.00	0.00	0.00
8	HSDPA	WiFi	MPEG-4	MPEG-4	1536.00	1536.00	0.00	0.00
9	HSDPA	WiFi	H.264	H.264	256.00	256.00	0.00	0.00
10	HSDPA	WiFi	H.264	H.264	1536.00	1536.00	0.00	0.00
11	WiFi	WiFi	H.264	MPEG-4	1536.00	1536.00	1.00	1.00
12	WiFi	WiFi	H.264	H.264	1536.00	256.00	1.00	1.00
13	WiFi	WiFi	H.264	MPEG-4	1536.00	1536.00	3.00	3.00
14	WiFi	WiFi	H.264	H.264	1536.00	256.00	3.00	3.00
15	WiFi	WiFi	H.264	MPEG-4	1536.00	1536.00	5.00	5.00
16	WiFi	WiFi	H.264	H.264	1536.00	256.00	5.00	5.00
17	WiFi	HSDPA	H.264	H.264	1536.00	1536.00	1.00	0.00
18	WiFi	HSDPA	H.264	H.264	1536.00	1536.00	3.00	0.00
19	WiFi	HSDPA	H.264	H.264	1536.00	1536.00	5.00	0.00
20	WiFi	WiFi	H.264	H.264	1536.00	1536.00	0.00	0.00
21	WiFi	WiFi	H.264	H.264	1536.00	1536.00	1.00	1.00
22	WiFi	WiFi	H.264	H.264	1536.00	1536.00	3.00	3.00
23	WiFi	WiFi	H.264	H.264	1536.00	1536.00	5.00	5.00
24	WiFi	WiFi	H.264	H.264	768.00	768.00	0.00	0.00
25	WiFi	WiFi	H.264	H.264	256.00	256.00	0.00	0.00
26	WiFi	WiFi	MPEG-4	MPEG-4	1536.00	1536.00	0.00	0.00
27	WiFi	WiFi	MPEG-4	MPEG-4	256.00	256.00	0.00	0.00
28	HSDPA	HSDPA	H.264	H.264	1536.00	1536.00	0.00	0.00
29	HSDPA	HSDPA	H.264	H.264	768.00	768.00	0.00	0.00
30	HSDPA	HSDPA	MPEG-4	MPEG-4	1536.00	1536.00	0.00	0.00
31	HSDPA	HSDPA	MPEG-4	MPEG-4	768.00	768.00	0.00	0.00
32	LAN	LAN	H.264	H.264	1536.00	1536.00	0.00	0.00
33	LAN	LAN	H.264	H.264	768.00	768.00	3.00	3.00
34	LAN	LAN	H.264	H.264	256.00	256.00	0.00	0.00
35	WiFi	WiFi	H.264	MPEG-4	256.00	256.00	0.00	0.00
36	WiFi	WiFi	H.264	MPEG-4	768.00	768.00	0.00	0.00
37	WiFi	WiFi	H.264	MPEG-4	1536.00	1536.00	0.00	0.00
38	WiFi	WiFi	MPEG-4	H.264	256.00	256.00	0.00	0.00
39	WiFi	WiFi	MPEG-4	H.264	768.00	768.00	0.00	0.00
40	WiFi	WiFi	MPEG-4	H.264	1536.00	1536.00	0.00	0.00
41	WiFi	WiFi	H.264	H.264	256.00	768.00	0.00	0.00
42	WiFi	WiFi	H.264	H.264	768.00	1536.00	0.00	0.00
43	WiFi	WiFi	H.264	H.264	1536.00	768.00	0.00	0.00

(Continued)

Table A.17 (Continued)

	$Network_1$	$Network_2$	$VCodec_1$	$VCodec_2$	$VBitrate_1$	$VBitrate_2$	$Loss_1$	$Loss_2$
44	WiFi	WiFi	H.264	H.264	768.00	256.00	0.00	0.00
45	WiFi	WiFi	MPEG-4	MPEG-4	256.00	768.00	0.00	0.00
46	WiFi	WiFi	MPEG-4	MPEG-4	768.00	1536.00	0.00	0.00
47	WiFi	WiFi	MPEG-4	MPEG-4	1536.00	768.00	0.00	0.00
48	WiFi	WiFi	MPEG-4	MPEG-4	768.00	256.00	0.00	0.00
49	WiFi	WiFi	H.264	H.264	256.00	1536.00	0.00	0.00
50	WiFi	WiFi	H.264	H.264	1536.00	256.00	0.00	0.00

Table A.18 Quality scores in Test V1

	MOS_v	CI_{up}	CI_{low}	G.1070	J.341	J.144
1	2.17	2.28	2.07	1.99	1.77	2.97
2	3.07	3.21	2.93	2.94	2.87	3.60
3	3.38	3.55	3.22	3.78	3.29	3.69
4	2.53	2.66	2.40	2.77	2.17	2.90
5	3.18	3.32	3.04	3.60	2.98	3.55
6	2.61	2.73	2.49	3.36	3.31	3.82
7	2.15	2.28	2.03	1.99	1.77	3.05
8	3.59	3.77	3.42	3.42	3.36	4.19
9	2.61	2.72	2.50	2.84	2.21	3.03
10	3.13	3.27	2.99	2.78	3.40	4.14
11	2.53	2.68	2.39	2.82	2.94	3.52
12	2.31	2.44	2.18	2.51	2.36	3.15
13	2.20	2.33	2.07	2.03	2.67	3.41
14	2.07	2.20	1.94	1.77	1.90	2.66
15	1.98	2.11	1.85	1.68	1.83	3.24
16	1.80	1.92	1.67	1.45	1.52	2.38
17	2.33	2.46	2.20	2.22	2.85	3.53
18	2.04	2.18	1.90	1.52	1.88	3.14
19	1.85	1.97	1.72	1.56	2.09	3.18
20	3.86	4.01	3.71	4.16	3.57	4.40
21	2.15	2.27	2.03	2.13	2.56	3.76
22	1.77	1.88	1.66	1.17	1.72	3.25
23	1.50	1.59	1.41	1.02	1.21	2.70
24	3.77	3.90	3.64	3.94	3.10	4.08
25	2.68	2.83	2.54	2.98	2.21	2.99
26	3.80	3.97	3.63	4.10	3.36	4.30
27	2.19	2.30	2.08	2.00	1.76	2.98
28	2.72	2.90	2.54	2.98	2.88	3.76
29	3.00	3.16	2.83	3.59	2.98	3.56
30	3.62	3.77	3.47	4.10	3.32	3.83
31	3.23	3.40	3.07	3.16	2.90	3.73
32	3.97	4.13	3.81	4.16	3.58	4.47
33	2.06	2.18	1.94	1.87	2.22	3.40
34	2.65	2.80	2.50	2.98	2.21	3.00
35	2.27	2.38	2.16	2.49	1.93	2.77
36	3.29	3.45	3.13	3.55	3.05	3.55
37	3.74	3.88	3.60	4.13	3.49	3.62
38	2.51	2.64	2.38	2.49	1.94	2.71
39	3.50	3.66	3.34	3.55	3.04	3.82
40	3.46	3.62	3.29	4.13	3.44	3.72
41	3.00	3.15	2.86	3.46	2.63	3.10
42	3.52	3.68	3.36	4.05	3.33	3.76
43	3.60	3.75	3.44	3.65	3.35	3.69
44	2.86	2.99	2.72	3.46	2.51	3.04
45	2.67	2.81	2.53	2.58	2.41	3.17

(Continued)

Table A.18 (Continued)

	MOS_v	CI_{up}	CI_{low}	G.1070	J.341	J.144
46	3.44	3.60	3.28	3.63	3.21	3.67
47	3.39	3.54	3.24	3.63	3.19	3.62
48	2.46	2.59	2.34	2.58	2.28	3.04
49	3.13	3.28	2.98	3.57	2.85	3.32
50	2.90	3.05	2.76	3.57	2.73	3.15

Table A.19 Description of conditions in Test V2, which includes the network, speech and video codec, video encoding bit rate, and packet loss configuration before and after the switch. If the corresponding values are equal, no switch was performed

	$Network_1$	$Network_2$	$ACodec_1$	$ACodec_2$	$VCodec_1$	$VCodec_2$	$VBitrate_1$	$VBitrate_2$	$Loss_1$	$Loss_2$
1	WiFi	WiFi	G.722.2	G.722.2	H.264	H.264	1536.00	1536.00	0.00	0.00
2	WiFi	WiFi	G.722.2	G.722.2	H.264	H.264	1536.00	1536.00	1.00	1.00
3	WiFi	WiFi	G.722.2	G.722.2	H.264	H.264	1536.00	1536.00	3.00	3.00
4	WiFi	WiFi	G.722.2	G.722.2	H.264	H.264	1536.00	1536.00	5.00	5.00
5	WiFi	WiFi	G.722.2	G.722.2	H.264	H.264	256.00	256.00	0.00	0.00
6	WiFi	WiFi	G.722.2	G.722.2	H.264	H.264	256.00	256.00	3.00	3.00
7	WiFi	WiFi	G.722.2	G.722.2	H.264	H.264	256.00	256.00	5.00	5.00
8	WiFi	WiFi	G.722.2	G.722.2	MPEG-4	MPEG-4	1536.00	1536.00	0.00	0.00
9	WiFi	WiFi	G.722.2	G.722.2	MPEG-4	MPEG-4	1536.00	1536.00	3.00	3.00
10	WiFi	WiFi	G.722.2	G.722.2	MPEG-4	MPEG-4	1536.00	1536.00	5.00	5.00
11	WiFi	WiFi	G.722.2	G.722.2	MPEG-4	MPEG-4	256.00	256.00	0.00	0.00
12	HSDPA	HSDPA	G.722.2	G.722.2	H.264	H.264	1536.00	1536.00	0.00	0.00
13	HSDPA	HSDPA	G.722.2	G.722.2	H.264	H.264	256.00	256.00	0.00	0.00
14	HSDPA	HSDPA	G.722.2	G.722.2	MPEG-4	MPEG-4	1536.00	1536.00	0.00	0.00
15	HSDPA	HSDPA	G.722.2	G.722.2	MPEG-4	MPEG-4	256.00	256.00	0.00	0.00
16	WiFi	HSDPA	G.722.2	G.722.2	H.264	H.264	1536.00	1536.00	0.00	0.00
17	WiFi	HSDPA	G.722.2	G.722.2	H.264	H.264	256.00	256.00	0.00	0.00
18	WiFi	HSDPA	G.722.2	G.722.2	MPEG-4	MPEG-4	1536.00	1536.00	0.00	0.00
19	WiFi	HSDPA	G.722.2	G.722.2	MPEG-4	MPEG-4	256.00	256.00	0.00	0.00
20	WiFi	WiFi	G.722.2	G.722.2	H.264	MPEG-4	1536.00	1536.00	0.00	0.00
21	WiFi	WiFi	G.722.2	G.722.2	H.264	MPEG-4	256.00	256.00	0.00	0.00
22	WiFi	WiFi	G.722.2	G.722.2	MPEG-4	H.264	1536.00	1536.00	0.00	0.00
23	WiFi	WiFi	G.722.2	G.722.2	MPEG-4	H.264	256.00	256.00	0.00	0.00
24	WiFi	WiFi	G.722.2	G.722.2	H.264	H.264	256.00	1536.00	0.00	0.00
25	WiFi	WiFi	G.722.2	G.722.2	H.264	H.264	1536.00	256.00	0.00	0.00
26	WiFi	WiFi	G.722.2	G.722.2	MPEG-4	MPEG-4	256.00	1536.00	0.00	0.00
27	WiFi	WiFi	G.722.2	G.722.2	MPEG-4	MPEG-4	1536.00	256.00	0.00	0.00
28	WiFi	HSDPA	G.722.2	G.722.2	H.264	H.264	1536.00	1536.00	1.00	0.00
29	WiFi	HSDPA	G.722.2	G.722.2	H.264	H.264	1536.00	1536.00	5.00	0.00
30	WiFi	WiFi	G.722.2	G.722.2	H.264	MPEG-4	1536.00	1536.00	1.00	1.00
31	WiFi	WiFi	G.722.2	G.722.2	H.264	MPEG-4	1536.00	1536.00	5.00	5.00
32	WiFi	WiFi	G.722.2	G.722.2	H.264	H.264	1536.00	256.00	1.00	1.00
33	WiFi	WiFi	G.722.2	G.722.2	H.264	H.264	1536.00	256.00	5.00	5.00
34	WiFi	WiFi	3GPP AMR	3GPP AMR	H.264	H.264	1536.00	1536.00	0.00	0.00
35	WiFi	WiFi	3GPP AMR	3GPP AMR	H.264	H.264	1536.00	1536.00	1.00	1.00
36	WiFi	WiFi	3GPP AMR	3GPP AMR	H.264	H.264	1536.00	1536.00	5.00	5.00
37	WiFi	WiFi	3GPP AMR	3GPP AMR	H.264	H.264	256.00	256.00	0.00	0.00
38	WiFi	WiFi	G.722.2	3GPP AMR	H.264	H.264	1536.00	1536.00	0.00	0.00
39	WiFi	WiFi	3GPP AMR	G.722.2	H.264	H.264	1536.00	1536.00	0.00	0.00
40	WiFi	WiFi	G.722.2	3GPP AMR	H.264	H.264	256.00	256.00	0.00	0.00
41	WiFi	WiFi	3GPP AMR	G.722.2	H.264	H.264	256.00	256.00	0.00	0.00
42	WiFi	HSDPA	3GPP AMR	3GPP AMR	H.264	H.264	1536.00	1536.00	1.00	0.00
43	WiFi	HSDPA	3GPP AMR	3GPP AMR	H.264	H.264	1536.00	1536.00	5.00	0.00

(Continued)

Table A.19 (continued)

	$Network_1$	$Network_2$	$ACodec_1$	$ACodec_2$	$VCodec_1$	$VCodec_2$	$VBitrate_1$	$VBitrate_2$	$Loss_1$	$Loss_2$
44	WiFi	WiFi	3GPP AMR	3GPP AMR	H.264	MPEG-4	1536.00	1536.00	1.00	1.00
45	WiFi	WiFi	3GPP AMR	3GPP AMR	H.264	MPEG-4	1536.00	1536.00	5.00	5.00
46	WiFi	WiFi	3GPP AMR	3GPP AMR	H.264	H.264	1536.00	256.00	1.00	1.00
47	WiFi	WiFi	3GPP AMR	3GPP AMR	H.264	H.264	1536.00	256.00	5.00	5.00
48	WiFi	WiFi	G.722.2	G.722.2	H.264	H.264	256.00	1536.00	3.00	3.00
49	WiFi	HSDPA	G.722.2	G.722.2	H.264	H.264	1536.00	1536.00	3.00	0.00
50	WiFi	WiFi	G.722.2	G.722.2	H.264	MPEG-4	1536.00	1536.00	3.00	3.00

Table A.20 Quality scores in Test V2

	MOS_{av}	$MOS_{a\|av}$	$MOS_{v\|av}$	E-model	POLQA	PESQ	G.1070	J.341	J.144	$MOS_{int\|av}$	Parametric	Signal-based
1	3.77	3.56	4.00	4.40	4.19	3.77	4.12	3.99	1.39	3.69	4.19	3.79
2	2.64	3.02	2.41	3.98	3.72	2.93	2.31	3.17	2.63	2.57	2.86	2.94
3	2.52	2.95	2.14	3.39	3.52	2.68	1.30	2.38	2.71	2.40	2.07	2.43
4	2.27	2.69	1.84	3.04	3.14	2.04	1.02	1.41	3.84	2.16	1.83	1.76
5	3.04	3.35	2.80	4.42	4.35	3.86	2.98	2.85	2.79	2.91	3.45	3.14
6	2.69	3.05	2.36	3.35	3.62	2.94	2.28	2.44	3.05	2.55	2.61	2.55
7	2.53	2.83	2.21	3.16	3.26	2.54	2.06	2.56	3.05	2.40	2.43	2.46
8	3.41	3.43	3.43	4.42	4.39	3.92	4.10	3.73	1.63	3.30	4.19	3.71
9	2.82	3.02	2.60	3.38	3.62	2.80	2.69	3.52	1.90	2.67	2.85	3.06
10	2.94	3.26	2.51	3.13	3.37	2.60	2.17	3.05	2.21	2.71	2.48	2.73
11	2.72	3.37	2.24	4.42	4.43	3.90	2.00	2.25	3.07	2.60	2.81	2.79
12	3.52	3.46	3.65	4.42	4.38	3.66	4.16	3.96	1.66	3.44	4.23	3.72
13	3.12	3.43	2.91	4.42	4.39	3.90	2.98	2.88	2.76	3.01	3.45	3.18
14	3.40	3.40	3.39	4.42	3.75	3.29	4.10	3.75	1.67	3.27	4.19	3.41
15	2.72	3.34	2.20	4.42	4.39	3.80	2.00	2.09	3.04	2.57	2.81	2.66
16	3.10	3.14	3.16	4.34	3.58	2.91	3.58	3.67	2.18	3.02	3.81	3.19
17	3.00	3.14	2.81	4.34	3.94	3.42	2.81	3.04	2.67	2.83	3.31	3.07
18	3.15	3.13	3.14	4.31	3.26	2.76	3.52	3.02	2.47	3.01	3.75	2.78
19	2.60	3.16	2.14	4.35	3.98	3.29	2.00	2.17	3.07	2.47	2.78	2.53
20	3.47	3.53	3.49	4.42	4.29	3.78	4.13	3.90	1.78	3.38	4.21	3.74
21	2.70	3.18	2.22	4.42	3.71	3.26	2.49	1.92	3.47	2.52	3.13	2.38
22	3.51	3.47	3.56	4.21	3.53	3.41	4.10	3.89	2.63	3.39	4.08	3.55
23	2.88	3.33	2.46	4.40	4.32	3.69	2.49	2.43	3.22	2.71	3.12	2.82
24	3.29	3.32	3.26	4.30	3.72	3.40	3.53	3.62	2.79	3.15	3.76	3.39
25	3.10	3.39	2.80	4.42	4.37	3.75	3.57	3.28	2.45	2.92	3.84	3.36
26	3.02	3.33	2.74	4.42	4.41	3.79	3.05	2.98	2.58	2.87	3.50	3.19
27	2.84	3.49	2.40	4.42	4.33	3.82	3.05	2.83	2.77	2.73	3.50	3.11

Table A.20 Quality scores in Test V2

	MOS_{av}	$MOS_{a\|av}$	$MOS_{v\|av}$	E-model	POLQA	PESQ	G.1070	J.341	J.144	$MOS_{int\|av}$	Parametric	Signal-based
28	2.48	2.54	2.39	3.64	2.82	1.94	2.38	3.23	2.58	2.38	2.78	2.53
29	2.61	3.07	2.22	3.85	3.50	2.60	1.96	2.55	2.60	2.49	2.59	2.48
30	2.99	3.35	2.72	3.92	3.91	3.15	2.96	3.57	2.05	2.86	3.23	3.25
31	2.58	2.91	2.20	2.98	3.08	2.31	1.63	2.31	2.82	2.42	2.14	2.26
32	2.71	3.11	2.42	3.99	4.11	3.23	2.66	2.83	2.99	2.60	3.08	2.88
33	2.35	2.85	1.96	3.10	3.38	2.50	1.46	1.64	3.50	2.27	2.08	2.00
34	3.46	2.80	4.02	3.52	3.00	2.71	4.16	4.06	1.35	3.31	3.76	3.29
35	2.62	2.60	2.41	3.27	2.87	2.43	2.27	3.46	2.11	2.41	2.58	2.86
36	2.02	2.37	1.62	2.93	2.43	1.95	1.02	1.34	3.60	1.96	1.80	1.71
37	2.83	2.62	2.82	3.52	2.81	2.44	2.98	2.81	2.81	2.63	3.08	2.55
38	3.26	2.66	4.01	3.01	2.47	1.49	4.16	4.03	1.40	3.24	3.50	2.65
39	3.25	2.76	3.86	3.01	2.69	1.69	3.93	4.04	1.59	3.22	3.37	2.76
40	2.81	2.58	2.71	3.01	2.76	1.60	2.98	2.81	2.83	2.56	2.87	2.21
41	2.80	2.63	2.82	3.01	2.65	1.76	2.98	2.85	2.78	2.63	2.87	2.29
42	2.56	2.28	2.62	3.17	2.06	1.81	3.04	2.67	2.95	2.40	2.96	2.24
43	2.49	2.43	2.36	3.19	2.48	2.13	1.77	2.65	2.58	2.33	2.28	2.35
44	2.76	2.61	2.69	3.30	2.81	2.53	2.78	3.53	2.36	2.56	2.87	2.94
45	2.26	2.35	2.04	2.85	2.39	1.87	1.63	2.07	2.97	2.15	2.10	2.00
46	2.56	2.50	2.43	3.41	2.46	2.07	2.47	2.23	3.33	2.39	2.75	2.14
47	2.15	2.34	1.86	2.85	2.06	1.72	1.51	1.63	3.83	2.06	2.04	1.77
48	2.42	2.78	2.07	3.39	3.69	2.89	1.72	2.40	2.93	2.30	2.31	2.51
49	2.25	2.38	2.13	2.87	1.83	1.53	2.28	2.39	2.95	2.20	2.45	2.02
50	2.71	2.94	2.43	3.36	3.40	2.53	2.14	2.90	2.77	2.55	2.54	2.63

Table A.21 Description of call quality profiles in Test V3, which includes the profile name, as well as the network, speech and video codec, video encoding bit rate, packet loss configuration, and the number of sample in the middle of which a parameter was switched

	Name	Network	ACodec	VCodec	Bitrate	Loss	Pos.
1	H.264 (high bit rate)	WiFi	G.722.2	H.264	1536	0	–
2	H.264 (high bit rate), 5% loss	WiFi	G.722.2	H.264	1536	5	–
3	H.264 (high bit rate) in HSDPA	HSDPA	G.722.2	H.264	1536	0	–
4	H.264 (low bit rate)	WiFi	G.722.2	H.264	256	0	0
5	Handover to HSDPA at high bit rate	WiFi to HSDPA	G.722.2	H.264	1536	0	3
6	Leaving WiFi (high bit rate)	WiFi	G.722.2	H.264	1536	0, 3, 3, 5, 5	0
7	Leaving WiFi, switch to HSDPA	WiFi to HSDPA	G.722.2	H.264	1536	0, 3, 3, 5, 5	3
8	Leaving WiFi, switch to MPEG4	WiFi	G.722.2	H.264 to MPEG-4	1536	0, 3, 3, 5, 5	3
9	Joining WiFi (low bit rate)	WiFi	G.722.2	H.264	256	5, 3, 3, 0, 0	0
10	Joining WiFi, switch to hi.	WiFi	G.722.2	H.264	256 to 1536	5, 3, 3, 0, 0	3
11	MPEG-4 (high bit rate)	WiFi	G.722.2	MPEG-4	1536	0	0
12	Quality improvement, beg.	WiFi	G.722.2	H.264	1536	0	1
13	Quality improvement, mid.	WiFi	G.722.2	H.264	1536	0	3
14	Quality improvement, end	WiFi	G.722.2	H.264	1536	0	5
15	Quality degradation, end	WiFi	G.722.2	H.264	1536	0	5
16	Strong degradation, mid.	WiFi	G.722.2	H.264	1536	0, 0, 5, 0, 0	-

Table A.22 Quality scores in Test V3

	Name	MOS_{av}	CI_{up}	CI_{low}	Mean	ETSI	Weiss	Rosenbluth	Signal-based	Parametric
1	H.264 (high bit rate)	3.71	3.91	3.52	3.77	3.77	3.77	3.77	4.08	4.19
2	H.264 (high bit rate), 5% loss	1.97	2.10	1.83	2.27	2.27	2.27	2.27	3.49	3.76
3	H.264 (high bit rate) in HSDPA	2.82	3.01	2.63	3.52	3.52	3.52	3.52	1.85	1.83
4	H.264 (low bit rate)	3.03	3.16	2.89	3.04	3.04	3.04	3.04	1.97	2.10
5	Handover to HSDPA at high bit rate	2.57	2.87	2.26	3.54	3.54	3.41	3.49	3.21	3.25
6	Leaving WiFi (high bit rate)	2.00	2.17	1.82	2.67	2.63	2.51	2.48	3.45	3.59
7	Leaving WiFi, switch to HSDPA	2.57	2.83	2.32	3.12	3.16	2.90	2.96	2.71	2.77
8	Leaving WiFi, switch to MPEG4	2.56	2.71	2.42	2.98	2.97	2.83	2.90	3.02	3.16
9	Joining WiFi (low bit rate)	2.66	2.81	2.51	2.80	2.82	2.74	2.83	1.99	2.12
10	Joining WiFi, switch to high	2.91	3.12	2.70	3.04	3.11	2.92	2.94	2.89	3.00
11	MPEG-4 (high bit rate)	3.01	3.16	2.86	3.08	3.11	3.03	3.08	2.41	2.56
12	Quality improvement, beg.	3.54	3.75	3.34	3.68	3.69	3.57	3.69	3.90	4.01
13	Quality improvement, mid.	3.20	3.38	3.03	3.38	3.42	3.32	3.38	3.05	3.18
14	Quality improvement, end	3.13	3.28	2.99	3.09	3.11	3.09	3.10	2.20	2.34
15	Quality degradation, end	3.46	3.65	3.26	3.64	3.59	3.42	3.55	3.92	4.10
16	Strong degradation, mid.	3.15	3.33	2.98	3.47	3.50	3.14	3.25	3.96	4.11

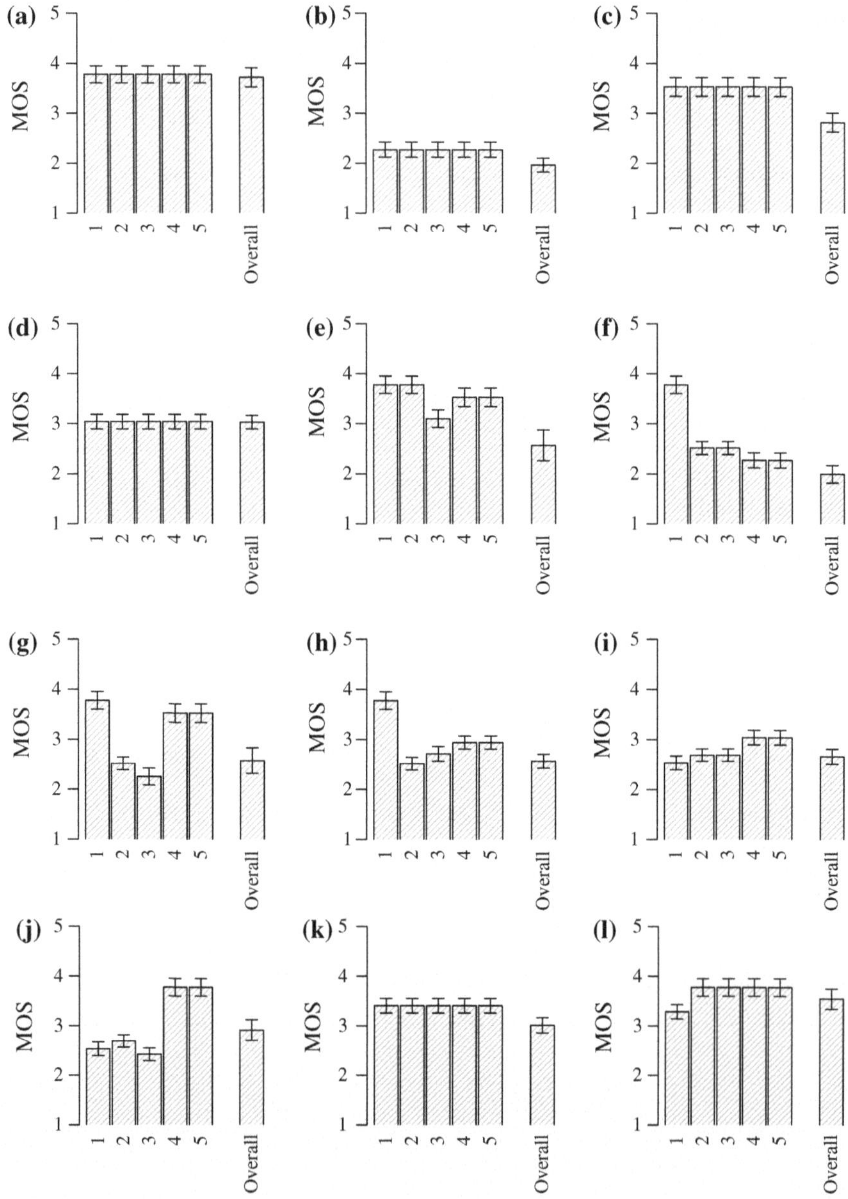

Fig. A.4 Call quality profiles of Test V3. MOS_{av} and 95% CI. **a** H.264 (high bit rate). **b** H.264 (high bit rate), 5% loss. **c** H.264 (high bit rate) in HSDPA. **d** H.264 (low bit rate). **e** Handover to HSDPA at high bit rate. **f** Leaving WiFi (high bit rate). **g** Leaving WiFi, switch to HSDPA. **h** Leaving WiFi, switch to MPEG-4. **i** Joining WiFi (low bit rate). **j** Joining WiFi, switch to high bit rate. **k** MPEG-4 (high bit rate). **l** Quality improvement, beg. **m** Quality improvement, mid. **n** Quality improvement, end. **o** Quality degradation, end. **p** Strong degradation, mid

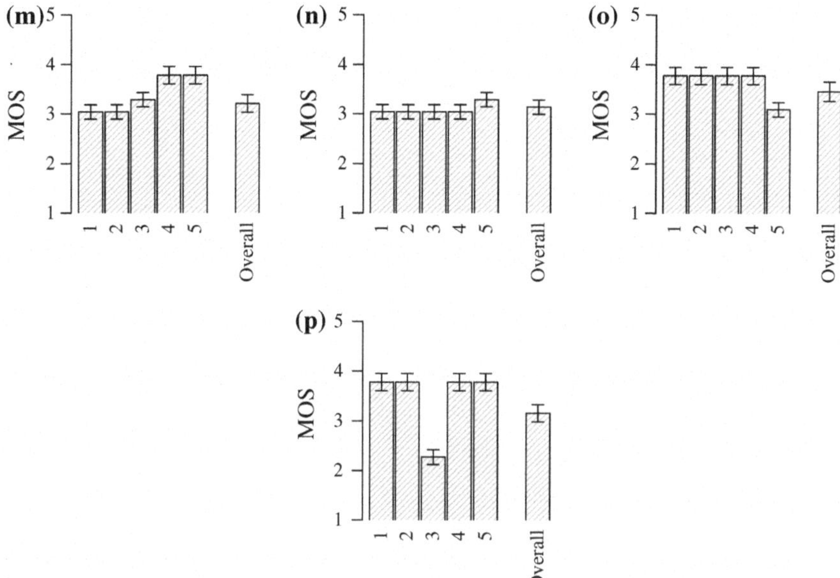

Fig. A.4 Continued

The manufacturer's authorised representative in the EU is Springer Nature Customer Service Centre GmbH, Europaplatz 3, 69115 Heidelberg, Germany. If you have any concerns regarding our products, please contact ProductSafety@springernature.com

Printed and bound by CPI Group (UK) Ltd, Croydon, CR0 4YY
15/07/2026
02167651-0001